INVENTORS and INVENTIONS

Volume 2

Communications – Fuller, R. Buckminster

Marshall Cavendish
Reference
New York

Marshall Cavendish
99 White Plains Road
Tarrytown, New York 10591-9001

www.marshallcavendish.us

Library of Congress Cataloging-in-Publication Data
Inventors and inventions.
     p. cm.
   Includes bibliographical references and index.
   ISBN 978-0-7614-7761-7 (set) -- ISBN 978-0-7614-7763-1 (v. 1) -- ISBN 978-0-7614-7764-8 (v. 2) -- ISBN 978-0-7614-7766-2 (v. 3) -- ISBN 978-0-7614-7767-9 (v. 4) -- ISBN 978-0-7614-7768-6 (v. 5) 1. Inventions--History. 2. Technology--History. 3. Inventors--History. I. Marshall Cavendish Corporation. II. Title.

   T15.I62 2007
   609--dc22

                                    2007060868

Printed in Malaysia (T)
11 10 09   3 4 5

Consulting Editor: Doris Simonis, Kent State University

Contributors: Richard Beatty; Jonathan Dore; Laura Lambert; Paul Schellinger; Mary Sisson; Gwendolyn Wells; Chris Woodford

MARSHALL CAVENDISH
Editor: Evelyn Ngeow
Publisher: Paul Bernabeo
Production Manager: Michael Esposito

MTM PUBLISHING
President: Valerie Tomaselli
Executive Editor: Hilary W. Poole
Editorial Coordinator: Tim Anderson
Editorial Assistants: Zachary Gajewski, Shalini Tripathi
Illustrator: Richard Garratt
Copyeditor: Carole Campbell
Design: Patrice Sheridan
Indexer: AEIOU, Inc.

Photographic Credits on page 640.

VOLUME 2

COMMUNICATIONS

Technology is usually designed to overcome human limitations; communications technology is no exception. Writing helps people to overcome the fallibility of memory and the loss of knowledge when others die; telephones carry information from place to place faster than a human ever could; and radio and television let one person talk to millions at a time.

Humans are social beings, so the problem of how to share information more effectively has always taxed inventors. A few thousand years ago, communication meant storing ideas and relating them to people in the immediate vicinity, using inventions such as the alphabet and written language. In the early 21st century, global communication is just as important and is facilitated by fiber optics and the Internet. Inventors and thinkers did not simply use science to move from the alphabet to the discoveries and technologies that permitted the development of the Internet; they both drove and were driven by wider changes in society.

FROM WRITING TO PRINTING

The earliest forms of writing were records of trade, kept in the Middle East around ten thousand years ago and consisting of clay tablets or bone. Around six thousand years later (ca. 1700 BCE), the Semites of the Mediterranean devised an alphabet. Dozens of inventions have

Internet bloggers work on their Web log stories during the 2004 Democratic National Convention in Boston, Massachusetts. In the 21st century, these online citizen journalists became a key part of the communications strategies of political campaigns.

since improved the way people store and share written language.

The first books were religious ones. An early book was not printed, like the books of today, but laboriously copied by hand onto a wax tablet called a codex. Later books were copied into large volumes made from parchment and vellum (durable writing materials produced from animal skins) bound in leather. In 105 CE Cai Lun (also known as Ts'ai Lun) invented paper based on tree bark, which was lighter, easier to produce, and less expensive. This enabled the Chinese to develop printing about four hundred years later. They laboriously carved an entire page of text into a large wooden block, covered it with ink, and pressed it onto the paper.

Another key advance took place around 1450 CE when a German, Johannes Gutenberg (ca. 1400–1468), invented printing with movable type. Instead of using one large printing block per page, Gutenberg used many tiny blocks, each capable of printing one let-ter, and he moved them around in a special frame, or form, so these separate blocks could be used again and again to print different words. Few people were literate in Europe in the Middle Ages, so most of the books printed on Gutenberg's press were religious works: priests and other officials of the Christian church were often literate and educated. The Bible was the first book Gutenberg printed, and around fifty copies of his version survive today. Once invented, Gutenberg's printing technology spread rapidly, carrying religious, political, and scientific information with it. By 1500, an estimated forty thousand books were in print in 14 European countries.

During the Industrial Revolution of the 19th century, the speed and volume of commerce increased, and old-fashioned handwriting, using pens dipped into inkwells, could no longer keep up. The 19th century brought new writing and printing technologies, including the steel-nibbed pen (1803), the graphite pencil

A Chinese woodcutter (left) and printer (right). To make a print, a piece of rice paper is pressed against the inked woodcut (silk painting, ca. 1800).

(1812), and raised-dot writing for the blind, invented in the 1820s by Louis Braille (1809–1852). Lewis Waterman (1837–1901), an insurance clerk, invented the fountain pen, which carried its own ink supply, in 1884; he wanted people to be able to sign his contracts more quickly. The even more convenient ballpoint pen, which applied thicker, quick-drying ink to paper using a tiny rolling ball, appeared in 1938, thanks to Hungarian Laszlo Biro (1899–1985).

The typewriter, developed by Christopher Latham Sholes (1819–1890) in 1868, was an even more far-reaching invention: it put the power of Gutenberg's movable-type printing press into the hands of individuals. With a typewriter, anyone could produce professional-looking printed materials. Quick and simple copying of documents, however, was not possible until 1938, when Chester Carlson (1906–1968) invented the photocopier.

Writing and printing technology allowed people to store and share their thoughts more easily and to pass on ideas to those who followed. The development of information-storing technologies, from written language to the photocopier, partly explains why the pace of invention is quicker now than it was thousands of years ago.

CARRYING MESSAGES, CARRYING SOUNDS

Initial improvements in communication came about largely through improvements in transportation: over the centuries, messages were relayed across long distances by everything from marathon runners and carrier pigeons in ancient Greece to stagecoaches, the Pony Express, and railroad locomotives in 19th-century America.

> You see, wire telegraph is a kind of very, very long cat. You pull his tail in New York and his head is meowing in Los Angeles. Do you understand this? And radio operates in exactly the same way: you send signals here, they receive them there. The only difference is that there is no cat.
>
> —Albert Einstein

All this changed with the growing scientific understanding of electricity and the realization that it could be harnessed for practical purposes. One of the most important discoveries was that electricity and magnetism were part of the same underlying phenomenon, electromagnetism, and either of them could cause the other. By 1837, English physicists Charles Wheatstone (1802–1875) and William Cooke (1806–1879) had sent simple electric messages down a cable, using magnetic compass needles to signal when the message arrived at the other end. This early electric telegraph was quickly superseded by a better system that artist Samuel Morse (1791–1872) developed in the United States in the 1840s. Like the transcontinental railroads that were laid across North America in the 19th century, the telegraph played a vital role in the westward expansion of the United States by helping the government to communicate rapidly with its new territories and control them effectively. By the end of the 19th century, the telegraph had spread worldwide, with countries linked by cables laid beneath the oceans. Communications technology was beginning to knit the world together.

Signaling systems, including the electric telegraph, were similar to printing inasmuch as they were forms of communication that individuals could not readily use themselves—they were too costly or cumbersome or they required special equipment and skilled operators. If a person wanted a pamphlet printed, it would have to be "set" in type and published by a skilled printer who owned a printing press; in the same way, if a person wanted to send a message from New York to Washington, D.C., a telegraph operator with the right equipment and trained in Morse code had to send it for that person. Just as the typewriter and photocopier put powerful printing technology in the hands of individuals, so the telephone, patented in 1876 by Alexander Graham Bell (1847–1922), accomplished the same end for long-distance communications.

By the dawn of the 21st century, many of the world's original copper telephone wires had been replaced by fiber-optic cables, which carry huge amounts of voice, fax, and Internet data as tiny pulses of laser light; Indian physicist Narinder Kapany (1927–) invented these remarkable "light pipes" in 1954 while he was studying in London. The fiber-optic revolution is an example of how social needs and technological developments constantly drive each other forward: high-capacity fiber-optic cables have made possible faster, cheaper international telephone calls and underpinned the growth of the global Internet; the demand for these services has been so great that scientists have been spurred on to develop even higher-capacity fiber-optic technologies.

Use of the telephone spread rapidly because it required no special skills, although for many decades trained operators were needed, first to connect all calls and later to handle long-distance and international calls. By the end of the 20th century, almost all calls could be made directly by individuals, who had only to pick up a handset and dial the correct series of numbers. Communication became more immediate, more direct, and less formal; unlike every previous form of long-distance communication, the telephone bypassed the need for written language and carried people's immediate thoughts, as well as their words, across geographic divides. When the telephone was first invented, the telegraph still had one major advantage: it could be connected to a hole-punching device that would

Stockbrokers at the trading firm of Tobey and Kirk study the ticker tape during the stock market crash in October 1929.

His Master's Voice, a painting of 1899 by Francis Berraud. Gramophone inventor Emile Berliner patented the image in 1900 to use as an advertisement for his company, Victor (later RCA/Victor).

automatically record incoming messages onto a reel of ticker tape that could be read later. The telegraph printer, with its famous ticker tape, revolutionized stock trading. It gave buyers and sellers the advantage of getting financial information more quickly and easily, greatly increased the volume of stock trading, and ultimately made Wall Street the major financial center in the United States.

The inventor of ticker tape, the prolific inventor Thomas Edison (1847–1931), quickly realized that something similar was needed for the telephone. In 1877, he invented the world's first telephone-answering machine—a rotating foil cylinder into which a vibrating needle dug grooves. German engineer Emile Berliner (1851–1929) realized the entertainment potential of this "talking machine," turned Edison's device into the gramophone, and launched the world's first record label (Victor) to market recordings for it. By 1948, gramophone records had become long-playing records (LPs) capable of storing roughly an hour of sound on two sides of a plastic disk. With the arrival of recorded music, singers and musicians became as important as the composers of music—their recordings, which would still be available after their deaths, potentially secured their fame. LPs evolved into compact discs during the late 1960s because of research by U.S. physicist James Russell (1931–).

Gramophones put prerecorded sound into the hands of ordinary people; the technology that allowed people to actually record sound was pioneered by another American inventor, Oberlin Smith (1840–1926). Building on Edison's work, Smith invented ways of recording sounds on cloth coated with magnetic chemicals. The German AEG Company commercialized this idea in 1935: its magnetophone could store sound by magnetizing a plastic tape covered with iron oxide that was wrapped into long reels. In the early 1950s, the American electrical engineer Charles Ginsburg (1920–1992) used similar technology to store television pictures, thus inventing the videotape recorder. During the 1960s, engineers at the Dutch Company Royal Philips Electronics squeezed huge reels of magnetic tape into tiny, convenient "compact cassettes" that could hold up to two hours of audio. Cassettes made recording sound easy, and when Akio Morita (1921–1999) invented the Sony Walkman in the 1980s, people could play music wherever they went. Compact audiocassettes revolutionized the way people listened to music, and a similar transformation turned Charles Ginsburg's cumbersome videotape reels into videocassettes, which also became popular in the 1980s.

RADIO, TELEVISION . . . AND THE TELEPHONE IS REBORN

Telegraphs and telephones connected the world by wires from the middle of the 19th century onward, but their big drawback was that they allowed communication only between fixed points. Nowhere was this problem more acute than on ships, one of the most important military technologies of the time. As business and commerce entered the telecommunications age, the battleships that waged wars and conquered nations seemed stuck behind in an earlier epoch: to send messages between one another and to the shore, they used primitive and unreliable methods like flashing lights and semaphore flags—useless over long distances and in bad weather.

Radio—a way of sending information through the air as a pattern of electricity and magnetism—revolutionized shipping, especially in military operations. The young Italian who developed radio, Guglielmo Marconi (1874–1937), failed his university entrance exam. Nevertheless, after studying research by

During World War II, two Navajo code-talkers in the Pacific relay orders over their field radio. The Navajo language was used as a code because it had no written form and very few speakers.

Communication–Then and Now

Imagine what communications technology would have been like in the 19th century. Most people lived near their immediate family and friends, so people would have had little need to send messages long distances: they spoke to one another in person or carried things on foot or horseback instead. For reaching more distant contacts, such as family members who had migrated to become pioneer settlers in the West, there was always the mail. The U.S. Post Office was created in 1789 and, although originally the country had only 75 post offices, the service rapidly expanded, following rivers and—much later—railroads into the new territories. Initially, mail took more than a month to travel from Washington, D.C., to Nebraska by stagecoach. American Express, which was originally a stagecoach company, was founded in 1850; Wells Fargo followed in 1852.

Sending mail between Missouri and California was slightly quicker if the correspondents used the Pony Express, established in 1860. This glamorous mail service advertised for "young, skinny, wiry fellows, not over 18, expert riders, willing to risk death daily, orphans preferred" to carry small packets in the saddlebags of swift horses. Pony Express was the e-mail of its day but was rendered obsolete after only 18 months by the arrival of the telegraph.

The great transportation and communications developments of the 19th century—railroads, telegraph, and the telephone—allowed people to keep in touch even as they traveled ever farther. These remarkable inventions allowed societies to reach outward and created, in their turn, the need for even better forms of communication. They also allowed national and eventually global markets to expand so greatly that people no longer think much about where goods and produce originate or how they reach their markets (in the early 19th century, almost all business and commerce was local). In the 21st century, with television pictures flooding in from around the world, people take global communications for granted. They can send instant messages over the Internet to friends on the other side of the world and get replies in seconds. In the 19th century, before telegraph and telephone cables were laid in the oceans, communicating with those in other countries could be done only by sending letters on ships, which could take months. A year or more could pass before a reply to a letter was received.

the greatest scientists of the day, he developed one of the most influential technologies of the 19th and 20th centuries. Radio brought new flexibility and speed-of-sound transmission to long-distance communications and also brought the ability to send messages quickly from one broadcaster to many listeners. Thus, it became an important technology for spreading news and propaganda (political messages). Radio played a crucial role in helping the Nazis to conquer and control other nations in the period leading up to World War II; the airwaves became one of the most important wartime battlegrounds.

Radio waves had uses other than carrying speech signals. Around the time Marconi was standing on the coast of England looking toward Canada, American engineer Reginald Fessenden (1866–1932) developed the first "wireless" telephone—it sent and received calls by radio waves. Popularized by the emergency services, the military, and taxi operators, radiotelephones gradually evolved into cellular phones; in their modern form, cellular phones were patented in 1973 by electrical engineer Martin Cooper (1928–). Radio was also the underlying technology that led engineers to devise ways of broadcasting pictures through the air—thus television was born. Numerous inventors were involved in developing television cameras (which converted pictures into electrical signals) and television sets (which turned the electrical signals back into pictures), but modern, electronic color television sets were actually pioneered by Hungarian-born American engineer Peter Goldmark (1906–1977) in the 1930s. Not all the important advances were made by scientists. The way in which cellphones and other wireless technologies share different radio frequencies owes much to a system called spread-spectrum communications, originally developed for military use by 1940s movie star Hedy Lamarr (1913–2000).

A cartoon from 1890 depicts a crowd inside a camera obscura, secretly viewing a couple on the beach outside. The caption reads, "Ah, Alicia, at last we are by ourselves, far from unsympathetic and prying eyes!"

STILL PHOTOGRAPHS AND MOTION PICTURES

During the 19th century, scientists not only learned how to record sound for posterity but also discovered how to "store" light with the development of photography. As long ago as the fourth century BCE, the Chinese developed a way of making images that later became known as a camera obscura. (The camera obscura was a darkened room with pinholes or slits poked in drapes over the windows to cast images onto an opposite wall). Although they could create images, they had no way of storing them.

Such storage technology was born in 1827 when French physicist Joseph Niépce (1765–1833) found he could record an image on a metal plate covered with bitumen (a kind of thick, black tar). This rather crude technique, which took eight hours to make a single photograph, was soon superseded by a better method called the daguerreotype, after the French artist who worked closely with Niépce to develop it, Louis Daguerre (1789–1851). William Henry Fox Talbot (1800–1877), an Englishman, improved photography further when he developed the negative–positive process still used today. In his method, paper coated with silver-based chemicals was exposed to light to produce a "negative" image of the scene being photographed, with light areas looking dark and vice versa; this negative was then "developed" (placed in various chemical baths) and used to print the scene in the "positive." Others substantially improved on Talbot's method in the years that followed, making sharper pictures that could be taken in much less time, more easily, and with less mess. Ultimately, in 1883, an American bank clerk, George Eastman (1854–1932), worked out a way of capturing photographs on plastic coated with light-sensitive chemicals. So the modern "film" was born; after Eastman invented his portable Kodak camera five years later, photography would soon become a hobby that almost anyone could enjoy. It was another example of technology gradually moving from scientists, specialists, and inventors into the hands of ordinary people.

Almost all people have desired entertainment—and photography provided that, too, in the form of motion pictures.

TIME LINE

ca. 8000 BCE	ca. 1700 BCE	105 CE	ca. 600	1450	1500	1820	1827	1837
Middle Eastern traders keep records on clay tablets or bone.	The Semites of the Mediterranean devise an alphabet.	Ts'ai Lun invents paper.	Printing with wooden blocks is developed in China.	Johannes Gutenberg invents printing with movable type.	An estimated forty thousand books are in print in Europe.	Louis Braille creates raised-dot printing for the blind.	Joseph Niépce and Louis Daguerre create the daguerreotype.	In the United Kingdom, Charles Wheatstone and William Cooke create an early electric telegraph.

Movies were born in 1893 when Thomas Edison's assistant, William Dickson (1860–1937), developed a mechanism that could pull a long reel of film through a camera called a kinetograph and take rapid-sequence photographs as it did so. When these many individual photographs, or frames, were later viewed in sequence at high speed in a matching film viewer or kinetoscope, the viewer's brain was tricked into seeing them as a single moving picture. Only one person at a time could view the kinetoscope, but it inspired two French brothers, Auguste Lumière (1862–1954) and Louis Lumière (1864–1948), to develop a better instrument, one that could project a film onto a wall for many people to view simultaneously. In 1894, the Lumières invented the cinématographe (movie projector) and, the following year, opened the world's first movie theater. Telegraphs and telephones may have changed the speed and convenience of communication, but photographs and movies changed its nature: they allowed people to communicate and share human emotions in a way that had never previously been possible.

THE DIGITAL WORLD, TODAY AND TOMORROW

All of these major strands in communications technology—from written language and books to radio and motion pictures—started life as separate inventions: written language was distinctly separate from spoken language, and pictures were different from words. From the mid-20th century onward, the spread of digital technology (a type of computerized electronics that processes information in the form of numbers) made possible the convergence of all these forms of information technology.

A key development was the user-friendly personal computer, which first appeared in the mid-1970s, pioneered by, among others, Steve Wozniak (1950–) and Steve Jobs (1955–), the founders of Apple Computer. Based on digital technology, personal computers gave individuals the power to store large volumes of information, edit it quickly and easily, and send it rapidly to others anywhere in the world. No other invention in history has had so much impact on the volume of information people store, process, and exchange and the speed with which they do so.

TIME LINE (continued)

1837	1868	1876	1877	1883	1884	1887	1893	1894
Samuel Morse invents his telegraph in the United States.	Christopher Latham Sholes invents the typewriter.	Alexander Graham Bell patents the telephone.	Thomas Edison invents the telephone-answering machine.	George Eastman creates modern photographic film.	Lewis Waterman invents the fountain pen.	Emile Berliner invents the gramophone.	William Dickson invents the kinetograph.	The cinématographe (movie projector) is invented by Auguste Lumière and Louis Lumière.

Communication Facts

- Between 1920 and 1970, the average number of long-distance telephone conversations per day grew from 1.5 million to 26.8 million.

- In 1946, the first year the U.S. government tracked the data, the United States had 8,000 television sets.

- In 2000, daily newspapers in the United States totaled 1,480, with a circulation of more than fifty-five million.

- Also in 2000, publishers sold 279 million adult hardcover books and 234 million young adult and children's hardcover books.

Since the beginning of the 21st century, people have been able to take digital photographs with a cell phone and e-mail them to friends, who can edit the photographs on a laptop and print them with an ordinary computer printer or post them on the Web, the information-sharing part of the Internet invented in 1989 by Tim Berners-Lee (1955–). All the above is possible because people now largely generate, store, process, and exchange information in digital form. The corporate world has taken notice; for example, although it was founded as a computer company, Apple launched the iPod music player in 2001 and the iPhone in 2007. People can use the Internet to listen to the radio or watch television, to download music, or to send information by e-mail or as instant messages. In the future, perhaps, people will no longer differentiate between television, telephone, and the Internet. Information—and the means for sharing it—will be all that matters.

—Chris Woodford

TIME LINE

1895	1930s	1935	1938	1938	1941	1950s	1954	1960s
Guglielmo Marconi builds a working radio.	Peter Goldmark pioneers the modern, electronic color television set.	The AEG Company releases the magneto-phone, which can store sound.	Chester Carlson invents the photocopier.	The ballpoint pen is invented by Laszlo Biro.	Hedy Lamarr patents a spread-spectrum communications system.	Charles Ginsburg creates the videotape recorder.	Narinder Kapany invents fiber-optic cables.	Compact disc invented by James Russell.

Further Reading

Books

Bridgman, Roger. *Eyewitness: Technology*. New York: Dorling Kindersley, 1998.

Crowley, David, and Paul Heyer. *Communication in History: Technology, Culture, and Society*. Boston: Allyn and Bacon, 2002.

Fischer, Claude S. *America Calling: A Social History of the Telephone to 1940*. Berkeley: University of California Press, 1992.

Goldsmith, Mike. *Fantastic Future*. New York: Scholastic, 2004.

Marvin, Caroline. *When Old Technologies Were New: Thinking about Electric Communication in the Late Nineteenth Century*. New York: Oxford University Press, 1990.

Woodford, Chris. *Communication and Computers*. New York: Facts on File, 2004.

Web sites

IEEE Virtual Museum
   Information and activities about electricity, electronics, and modern communications technology.
   http://www.ieee-virtual-museum.org/

Explore Invention at the Lemelson Center
   Explores the history of invention and encourages young people to create new inventions.
   http://invention.smithsonian.org/home/

National Inventors Hall of Fame
   Profiles of many well-known inventors.
   http://www.invent.org/

See also: Bell, Alexander Graham; Berners-Lee, Tim; Biro, Laszlo; Braille, Louis; Caselli, Giovanni; Cooper, Martin; Daguerre, Louis; Eastman, George; Edison, Thomas; Ginsburg, Charles; Goldmark, Peter; Gutenberg, Johannes; Jobs, Steve, and Steve Wozniak; Kapany, Narinder; Lamarr, Hedy; Lumière, Auguste, and Louis Lumière; Marconi, Guglielmo; Morita, Akio; Morse, Samuel; Nesmith Graham, Bette; Sholes, Christopher Latham; Wheatstone, Charles.

TIME LINE (continued)

1960s	1973	1976	1980s	1989	2000	2007
Engineers at the Philips Company create the cassette tape.	Martin Cooper invents the modern cellular phone.	Steve Jobs and Steve Wozniak begin selling the Apple I computer.	The Walkman is invented by Akio Morita.	Tim Berners-Lee invents the World Wide Web.	There are 1,480 daily newspapers in the United States.	Apple launches the iPhone.

COMPUTERS

Computers are unlike some other inventions in that no single person invented them and the idea was not born at a definite moment in time. In fact, computers have been developing since the mid-seventeenth century and continue to evolve more than three centuries later. In their original form, computers were wood-and-metal calculators used for mathematics and science. By the mid-20th century, they had become giant electronic machines that occupied entire rooms. In the early 21st century many "computers" are not even recognizable as computers: they are the invisible "brains" built into such electronic devices as cell phones, MP3 music players, and televisions.

MECHANICAL CALCULATORS

The first calculator—the abacus—was also the simplest. Still popular in China, it consists of beads (representing numbers) that slide along wires, enabling people to carry out complex calculations both quickly and efficiently. It was developed in Babylonia (now the southeastern part of Iraq) around 500 BCE and remained in widespread use for more than two thousand years. The earliest challenge to the supremacy of the abacus came in 1642, when French mathematician and philosopher Blaise Pascal (1623–1662) made the first mechanical calculator. It had a series of gears to represent the units—tens, hundreds, and so on—of a decimal number. The gears interlocked and added or subtracted numbers as they rotated.

Several decades later, in the 1670s, German mathematician Gottfried Leibniz (1646–1716) invented a more advanced calculator. Also using gears, it worked in a

A hand-colored woodcut from Margarita Philosophica *(1503) by Gregor Reish shows arithmetic being done with a counter.*

way broadly similar to Pascal's machine, but it could also multiply, divide, and calculate square roots. Leibniz's calculator pioneered another important feature of modern computers: a temporary memory (register) for storing numbers during a calculation.

Another of Leibniz's significant inventions was a way of representing any decimal number using only the digits zero and one. For example, the decimal number 86 can be represented in binary as 1010110. More than two hundred years later, this simple idea would become the basis of how all computers stored information.

PROGRAMMING PIONEERS

Electronics, a way of controlling machines using electricity, was invented at the beginning of the 20th century; earlier calculators and computers were entirely mechanical (made from wheels, gears, levers, and so on). Calculators and computers differ in the amount of human direction they need to operate. Whereas a calculator merely carries out actions one by one at the direction of an individual, a computer can perform a sequence of operations with little or no human intervention. The series of instructions a computer follows is called a program.

Calculators began to evolve into computers during the early decades of the 19th century. To function as a computer, a calculator needed a mechanism that could store and carry out its programs. Such a device was invented in 1801 by Frenchman Joseph-Marie Jacquard (1752–1834) for controlling a loom. Jacquard's loom was "programmed" using little pieces of card punched with holes. The position of the holes indicated how complex pat-

A modern digital camera contains more powerful computer technology than the giant ENIAC machine dating from 1944. A typical cell phone contains faster computer chips than IBM's original 1981 personal computer.

terns were to be woven in rugs and other textiles.

Jacquard's loom inspired British mathematician Charles Babbage (1791–1871), who dreamed of constructing elaborate mechanical computers. Babbage's machines proved too expensive to build and none were completed during his lifetime. Nevertheless, he was the first person to discern how a programmable computer could work. Like all modern computers, his designs had an input (a punched card mechanism, for feeding in numbers), a processor (a complex array of more than fifty thousand gear wheels that did mathematical calculations), and an output (a printing mechanism for showing the results).

Among the first to make practical calculating machines, like the ones Babbage had envisaged, were American inventor William Seward Burroughs (1857–1898) and statistician Herman Hollerith (1860–1929). After developing a simple, mechanical calculator, Burroughs formed a company that soon became the biggest manufacturer of adding machines in the United States. The company later moved into making general office machines, such as typewriters and computers. Herman Hollerith's work led in a similar direction. While compiling data for the U.S. Census in

1880, he realized he could make a machine that would do the job for him. He later founded the company that in 1924 was renamed International Business Machines (IBM), a pioneer in the computer industry during the 20th century.

ELECTRONIC COMPUTERS

In the 1920s, a U.S. government scientist, Vannevar Bush (1890–1974), began building complex mechanical calculators known as analog computers. These unwieldy contraptions used rotating shafts, gears, belts, and levers to store numbers and carry out complex calculations. One of them, Bush's Differential Analyzer, was effectively a gigantic abacus, the size of a large room. It was used mainly for military calculations, including those needed to aim artillery shells.

Analog computers led to digital machines, in which numbers were stored electrically in binary form, the pattern of zeros and ones invented by Leibniz. A prototype binary computer was developed in 1939 by U.S. physicist John Atanasoff (1903–1995) and electrical engineer Clifford Berry (1918–1963). The first major digital computer—one that stored numbers electrically instead of representing them with wheels and belts—was the Harvard Mark I. It was completed at Harvard University in 1944 by mathematician Howard Aiken (1900–1973) and used 3,304 relays, or telephone switches, for storing and calculating numbers. Two years later, scientists at the University of Pennsylvania

A vacuum tube is displayed by Gordon Moore, cofounder of Intel Corporation.

built ENIAC, the world's first fully electronic computer (see box, The First Electronic Computer). Instead of relays, it used almost eighteen thousand vacuum tubes, also known as valves, which operated more quickly. According to its operating manual, "The speed of the ENIAC is at least 500 times as great as that of any other existing computing machine."

THE TRANSISTOR AGE

Switches, or devices that can be either "on" or "off," can be considered the brain cells of a computer: each one can represent a single binary—zero or one. The more switches a computer has, the more numbers it can store; the faster the switches flick on or off, the faster the computer operates. Vacuum tubes were

The First Electronic Computer

The modern computer age began in 1946 when John Mauchly (1907–1980) and J. Presper Eckert (1919–1995) of the University of Pennsylvania built a monstrous computer named the ENIAC (Electronic Numerical Integrator and Calculator). The ENIAC was an astonishing feat of engineering, the electronic equivalent of Isambard Kingdom Brunel's steamships or John Roebling's Brooklyn Bridge. It was approximately one hundred feet (30.5 m) long (the same as five cars parked fender to fender), eight feet (2.4 m) high, and weighed around thirty tons (27.2 metric tons, as much as five elephants). It had to be that big because it contained 17,468 vacuum tube switches, enough electrical cable to stretch from New York City to Detroit, and around five million hand-soldered electrical connections.

Despite its gargantuan dimensions, the ENIAC was roughly one million times slower than a fast modern PC and a thousand times more expensive. Harry Reed, one of the computer scientists who worked with the ENIAC, described it as "a very personal computer. Now we think of a personal computer as one that you carry around with you. The ENIAC was one that you kind of lived inside." The ENIAC was also much harder to use than a modern PC. It is often described as the world's first fully electronic, general-purpose computer, because in theory it could be programmed to do different jobs. In the ENIAC, programming required rewiring the entire machine to work a different way. That was an immensely complex and laborious process and, according to Harry Reed, all part of the challenge: "One was supposed to suffer to do it."

A technician changes one of the many thousands of vacuum tubes in the ENIAC.

faster switches than relays, but each one was the size and shape of an adult's thumb; eighteen thousand of them took up a huge amount of space. To work properly, vacuum tubes had to be permanent- ly heated to a high temperature, so they consumed enormous amounts of electricity. Computers were evolving quickly during the 1940s, largely driven by military needs. Yet the size, weight, and power consumption of vacuum tubes had become limitations. Putting a computer in an airplane or a missile was impossible; a better kind of switch was needed.

Toward the end of the 1940s, this problem was solved by three scientists at Bell Telephone Laboratories (Bell Labs) in New Jersey. John Bardeen (1908–1991), Walter Brattain (1902–1987), and William Shockley (1910–1989) were trying to develop an amplifier that would boost telephone signals so they could travel much farther. The device they invented in late 1947—the transistor—turned out to have a more important use in computer switches. A single transistor was about as big as a pea and used virtually no electricity, so computers made from thousands of transistors were smaller and more efficient than those made from vacuum tubes.

Originally, transistors had to be made one at a time and then hand-wired into complex circuits. Working independently, two American electrical engineers, Jack Kilby (1923–2005) and Robert Noyce (1927–1990), devised a better way of making transistors, called an integrated circuit, in 1958. Their idea enabled thousands of transistors—and the intricate connections between them—to be manufactured in miniaturized form on the surface of a piece of silicon, one of the chemical elements of sand.

COMMERCIAL COMPUTING

In 1943, before transistors were invented, IBM executive Thomas Watson, Sr., had reputedly quipped: "I think there's a world market for about five computers." Within a decade, his own company proved him wrong—using vacuum tubes to manufacture its first general-purpose computer, the IBM 701, for 20 different customers. The arrival of transistors changed everything, making companies such as IBM able to develop increasingly affordable business machines from the 1950s onward.

Until that decade, most of the advances in computing had come about through improvements in hardware—the mechanical, electrical, or electronic components from which computers are made. During the 1950s, advances were also made in software—the programs that control how computers operate. Much of the credit goes to Grace Murray Hopper (1906–1992). A mathematician who divided her time between the U.S.

TIME LINE

ca. 500 BCE	1642	1670s	1801	1820s–1871	1891
Abacus developed in Babylonia.	Blaise Pascal makes the first mechanical calculator.	Gottfried Leibniz invents a calculator with temporary memory and a method to store numbers using a binary system.	Joseph-Marie Jacquard invents programmable punch cards.	Charles Babbage describes how a programmable computer could work.	William Burroughs develops the first practical calculating machine.

Navy and several large computer corporations, Hopper is remembered as a pioneer of modern computer programming. She invented the first compiler, a computer program that "translates" English-like commands that people understand into binary numbers that computers can process, in the 1950s.

PERSONAL COMPUTING

Even the invention of the transistor did not make computers affordable enough for most people. The world's first fully transistorized computer, the PDP-1 made by Digital Equipment Corporation, still cost approximately $120,000 when it was launched in 1960. Toward the end of the 1960s, however, another breakthrough was made in electronics; it changed everything. Robert Noyce, coinventor of the integrated circuit, and Gordon Moore (1929–) formed a company called Intel in 1968. The following year, one of Intel's engineers, Marcian Edward (Ted) Hoff (1937–), developed a way of making a powerful kind of integrated circuit that contained all the essential components of a computer. It was the microprocessor, popularly known as a microchip or silicon chip.

Microchips were no bigger than a fingernail and, during the early 1970s, found their way into various electronic devices, including digital watches and pocket calculators. In the mid-1970s, electronics enthusiasts started using microprocessors to build their own home computers. In 1976, two California hobbyists, Steve Jobs (1955–) and Steve Wozniak (1950–), used this approach to develop the world's first easy-to-use personal "microcomputers": the Apple I and Apple II.

TIME LINE (continued)

1920s	1924	1939	1944	1946	1947
Vannevar Bush builds the first analog computers.	Herman Hollerith founds what will become IBM.	John Atanasoff and Clifford Berry develop the first binary computer.	Howard Aiken completes the first major digital computer, Harvard Mark I.	The first fully electronic computer, ENIAC, is completed.	John Bardeen, Walter Brattain, and William Shockley invent the transistor.

Women in Computing

When Charles Babbage was devising his early
mechanical computers, he was assisted by Augusta Ada
Byron King (1815–1852), daughter of the English poet,
Lord Byron. Many describe Ada King as the world's first computer programmer because she helped Babbage to realize that his
machine could be reprogrammed to do different jobs.

A century later, women were making pioneering contributions with
the first electronic computers. Before the ENIAC was invented, the army
used 75 young women who were known as "computers" to calculate its
missile-firing tables. Six of them were later responsible for operating and
programming the ENIAC. Adele Goldstein, wife of an ENIAC scientist,
wrote the machine's operating manual. When the ENIAC's inventors,
John Mauchly and J. Presper Eckert, formed their own corporation in the
late 1940s, one of their employees was Grace Murray Hopper. The inventor of the computer compiler, Hopper had programmed the Harvard Mark
I and written its technical manual. During the 1960s, perhaps the greatest achievement in computing was guiding Apollo space rockets to the
moon. Some of the important Apollo programs were written by Evelyn
Boyd Granville (1924–), an African American mathematician who had
been lent to NASA by her employer, IBM.

In modern times, women are as likely as men to make brilliant computer scientists or dot-com Internet pioneers. Working at Sun
Microsystems in the 1990s, Kim Polese managed the development of Java,
a powerful programming language designed for the World Wide Web.
Around the same time, in Japan, Mari Matsunaga helped to launch iMode,
a popular way of using the Internet on cell phones. For this achievement,
she was later singled out as Asia's top businesswoman by *Forbes* magazine.
Pat Sueltz began her career as a telephone linewoman in Los Angeles in the
mid-1970s. Twenty years later, she was technical adviser to IBM's chief
executive officer; she then became a president at Sun Microsystems, and
later chief executive officer of several major Internet firms.

Many women have been encouraged to pursue careers in computing as
a result of the efforts of American computer scientist Anita Borg
(1949–2003). She formed an online community called Systers in 1987,
which now has around 2,300 women members in 35 countries. Until her
death from brain cancer in 2003, Borg battled to change the attitude that
women "have to be different or strange to get into this field."

Many other companies launched microcomputers, all of them with incompatible hardware and software. Apple's machines were so successful that, by 1981, IBM was forced to launch its own personal computer (PC), using software developed by the young, largely unknown Bill Gates (1955–) and his then-tiny company, Microsoft. During the 1980s, most personal computers were standardized around IBM's design and Microsoft's software. This was extremely important for businesses in particular, as employees needed to be able to share information easily.

USER-FRIENDLY COMPUTERS

Users needed not just compatibility but also ease of operation. Most of the early computer programmers had been highly qualified mathematicians; ordinary people were less interested in how computers worked than in what they could actually do. One inventor who recognized this immediately was American computer scientist Douglas Engelbart (1925–). During the 1960s, he pioneered a series of inventions that made computers more "user-friendly," the best known of which is the computer mouse.

Engelbart's ideas were taken up in the early 1970s at the Palo Alto Research Center (PARC), a laboratory in California that was then a division of the Xerox Corporation. Xerox had made its name and fortune in the 1960s, manufacturing photocopiers invented by Chester Carlson (1906–1968) in 1938. Executives at Xerox believed the arrival of computers heralded a new "paperless" era in which photocopiers and the like might become obsolete. Consequently, Xerox began developing revolutionary, easy-to-use office computers to ensure that it could remain in business long into the future.

After visiting Xerox PARC, Apple's head, Steve Jobs, began a project to develop his own easy-to-use computer, initially called PITS (Person In The Street). This eventually evolved into Apple's popular Macintosh (Mac) computer, launched in 1984. When Microsoft incorporated similar ideas into its own Windows software, Apple was unable to stop the Redmond, Washington, company despite a lengthy court battle. The ultimate victors were computer users, who have seen an enormous improvement in the usability of computers since the mid-1980s.

MODERN COMPUTING

From the abacus to the PC, computers had been largely self-contained machines.

TIME LINE (continued)					
1952	**1958**	**1960s**	**1960**	**1968**	**1969**
Grace Murray Hopper publishes the first computer compiler.	Jack Kilby and Robert Noyce invent the integrated circuit.	Douglas Engelbart begins to develop user-friendly hardware.	The first fully transistorized computer, PDP-1, is created.	Robert Noyce and Gordon Moore establish Intel.	Marcian Edward Hoff develops the microprocessor.

How Did Computers Get So Small?

1   The first computers were powered by large vacuum tubes. Vacuum tubes are large—about the size of an adult thumb—and a single computer required thousands of them; the first fully electronic computer, ENIAC, used almost 18,000 vacuum tubes. The first computers often took up entire floors of buildings.

2   The second generation of computers were much smaller— about the size of a room— because scientists began using transistors to power them. Transistors were much smaller than vacuum tubes and required less energy.

3   The third generation of computers were even smaller due to the invention of the integrated circuit, which could fit thousands of transistors on one small silicon chip. This led to the invention of a computer that was small enough to fit on a desk.

4   The fourth generation of computers— the computers we use today—are the smallest and most powerful yet. This is due to the invention of the very large scale integrated (VLSI) circuit, which fits millions of transistors onto one small chip.

That began to change in the 1980s, when the arrival of standardized PCs made connecting computers into networks easier. Businesses, schools, universities, and home users found they could use networked computers to share information more easily than ever before. More and more people connected their computers, mainly using the public telephone system, to form what is now a gigantic worldwide network of computers called the Internet.

During the late 1980s, British computer scientist Tim Berners-Lee (1955–) pioneered an easy way of sharing information over the Internet that he named the World Wide Web. Since then, the Web has proved to be one of the most important communication technologies ever invented. Apart from information shar-

---

TIME LINE

1976	1981	1984	Late 1980s	1990s	1991	1995
Steve Jobs and Steve Wozniak develop the Apple I and Apple II.	IBM launches its first personal computer.	Apple releases the Macintosh computer.	Tim Berners-Lee invents the World Wide Web.	Kim Polese manages the development of the Java programming language.	Linus Torvalds creates the Linux operating system.	Pierre Omidyar establishes eBay.

Initially conceived as a music player, Apple iPods can also be used to watch movies and play games. Other, similar devices also include cellular phones.

ing, it has helped people create new businesses, such as the popular auction Web site eBay, founded by American entrepreneur Pierre Omidyar (1967–) in 1995. In addition to offering new business opportunities, the Internet is also helping computing to evolve. One notable example of this evolution is the Linux operating system, originally created by Finnish computer programmer Linus Torvalds (1969–) in 1991. This software was developed by thousands of volunteers working together over the Internet.

Collaboration via the Internet is one of the most important aspects of modern computing. Another is convergence: a gradual coming together of computers and other communication technologies. Telephones, cameras, televisions, computers, stereos, and sound recording equipment were once entirely separate. Now, all these technologies can be incorporated into a single pocket-size device such as a cell phone. Collaboration and convergence indicate that computer technology is continually evolving.

—Chris Woodford

Further Reading

Books

Cringely, Robert X. *Accidental Empires: How the Boys of Silicon Valley Make Their Millions, Battle Foreign Competition, and Still Can't Get a Date.* New York: HarperBusiness, 1996.

Williams, Brian. *Computers: Great Inventions.* Portsmouth, NH: Heinemann, 2001.

Woodford, Chris. *Communications and Computers.* New York: Facts On File, 2004.

Wurster, Christian. *Computers: An Illustrated History.* New York: Taschen, 2002.

Web sites

Computer History Museum
    A California museum whose Web site includes online exhibits, photographs, and a time line.
    http://www.computerhistory.org/

IEEE Virtual Museum
    The history of electricity, electronics, and computers.
    http://www.ieee-virtual-museum.org/

Triumph of the Nerds
    A PBS Web site about the development of personal computers in California in the 1980s.
    http://www.pbs.org/nerds/

See also: Babbage, Charles; Bardeen, John, Walter Brattain and William Shockley; Berners-Lee, Tim; Engelbart, Douglas; Hopper, Grace Murray; Jobs, Steve, and Steve Wozniak; Kilby, Jack, and Robert Noyce; McAfee, John; Omidyar, Pierre; Torvalds, Linus.

CONTESTS

The inventing contest is a method for turning promising ideas into winning inventions. Each year, dozens of contests provide showcases for all kinds of new creations, from the intensely practical to the bizarre. Typically offering large cash prizes and national publicity, contests are a great opportunity for budding inventors to test the worth of their ideas and potentially realize dreams of fame and fortune.

PUBLIC CHALLENGES

The earliest inventing contests were open challenges in which national governments offered large cash prizes to inventors who could solve problems that were important to society. One of the most famous was launched in 1714 by the British government, which offered a prize of £20,000 (equivalent to more than $1 million today) to an inventor who could solve "the longitude problem": find an accurate and reliable way for ships to work out their longitude (east-west position on the globe) while in the middle of the ocean. English watchmaker John Harrison (1693–1776) finally won the prize almost six decades later, after

making a series of chronometers (highly accurate clocks and watches) with which sailors could calculate their longitude.

Another historic contest took place in 1795, when the French government offered a prize of 12,000 francs (equivalent to $36,000 today) to anyone who could find a method of storing provisions for troops on the battlefield. The winner was Nicolas-François Appert (ca. 1750–1841), inventor of the modern method of canning food.

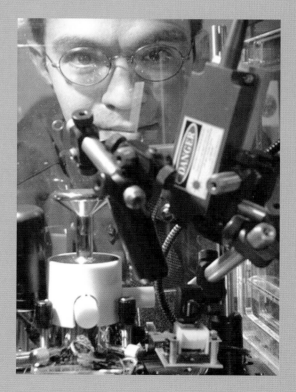

Graduate student Brian Hubert poses with his invention, the nano assembly machine, which was the winner of the MIT-Lemelson Student Prize for Inventiveness in 2001.

Although not a formal contest, a public challenge inspired one of the greatest accomplishments of the 20th century. In May 1961, President John F. Kennedy famously began the Apollo space program with a challenge to the American people: "I believe that this nation should commit itself to achieving the goal, before this decade is out, of landing a man on the moon and returning him safely to the earth." No cash prize was offered, and no single inventor completed the challenge: the program spent $25 billion and involved 400,000 people working closely together in perhaps the biggest example of collaborative invention of all time.

THE NEED FOR CONTESTS

Many people have ideas for new products but never get around to seeing whether the ideas are good enough to manufacture. Inventors may have to make dozens or even hundreds of prototypes (test versions) before they perfect their ideas. Most inventions take years to develop, few ever reach the marketplace, and even fewer are ultimately successful. With no instant reward, people have little incentive to invent. Contests with prizes and publicity are fun and give people a reason to turn ideas into a tangible device. Since most contests have entry deadlines, they focus people's minds, forcing them to develop their ideas in a way that might not happen otherwise. In short, contests act as a powerful spur to invention.

It is a common belief that invention means simply having a good idea, yet much else must happen before even the greatest inventions can make it into production. Inventions have to be manufac-

tured cost-effectively or they will never turn a profit. Many inventions have to comply with a range of safety laws and other stringent criteria. New products must fulfill a genuine need, stand out in an often crowded marketplace, and make people want to buy them. In a contest, with judges (often accomplished inventors themselves) scrutinizing the entries, all these factors are carefully considered. Expert advice from judges can help turn a vague idea into something with a greater promise for success.

INSPIRING INVENTORS

Feedback from experienced judges is just one example of how contests can give budding inventors support and encouragement. Many inventing contests also have their own Web sites, offering all kinds of useful advice and resources. When contests are held in public, inventors get the chance to meet one another and exchange ideas. Such interaction is part of the creative process that helps to inspire winning inventions. Even if inventors do not win a prize, the very experience of entering a contest may encourage them to try harder next time.

One of the most important aspects of contests is the way they raise the public profile of inventing. People sometimes dislike science at school, and media coverage can make technical subjects seem too remote from ordinary people's lives. As a result, the public does not always take an interest in cutting-edge technology. By contrast, inventions are seen as surprising, fun, entertaining, and educational. They may be extremely practical or entirely frivolous, but creative ideas always capture people's imagination. Thus, inventing contests help to get people more interested in science and technology, whether they take part as contestants or simply come to look. In short, most people simply enjoy the fun of a competition.

Contests cost money and are typically sponsored by well-known corporations such as 3M, Coca-Cola, Siemens, DuPont,

American Inventor

British entertainment producer Simon Cowell made his name in the United States with a successful television series called *American Idol*, a contest in which budding young pop stars compete for the chance to become rich and famous. In 2005, Cowell's company launched *American Inventor*, a television series along similar lines, for budding inventors. Though much less successful than its musical predecessor, *American Inventor* nevertheless managed to capture the public imagination by offering a $1 million first prize—twice as much as the prestigious Lemelson-MIT contest.

Cowell and his team had 10,000 applications from inventors eager to appear on the show. The number was gradually whittled down to just nine finalists. The inventors were given $50,000 each to develop their ideas, which they then had to pitch to a panel of judges and a studio audience. Having heard the judges' verdict, viewers then voted for the best idea. In the first series, the most promising inventions included a clip women could carry in their handbags to fasten broken restroom doors, a portable gym that could be folded into a bag, and an umbrella that collapses neatly into its own handle.

The 2005 series was won by Californian mechanical engineer Janusz Liberkowski. His triumphant invention was a new design of car chair called a spherical safety seat. Inspired by the tragic death of the inventor's young daughter in a car accident several years before, the seat protects children more effectively in an impact by holding them inside a series of spinning containers. Apart from the $1 million prize, Liberkowski earned the chance to have his invention developed by a leading manufacturer. The runner-up on *American Inventor* was former Chicago teacher Ed Hall, who developed an electronic game called "Word Ace" to improve children's spelling. His appearance on the show also led to an opportunity to work with toy manufacturers to move his idea toward production.

In 2004 the winner of the Staples Invention Quest was the WordLock, invented by Eric Gibbons.

Xerox, and Toshiba. While some fund contests purely out of goodwill, most see benefits from linking their names to innovation and creativity.

Contests may also be a source of new products. Each year, Wild Planet Toys challenges children to invent original toys, often putting the winning ideas into production. Another example is the Invention Quest run by the Staples office equipment firm, which sets out to find "the hottest office essential or a new, innovative organizational product" that makes work easier. Apart from a $25,000 cash prize, the contest offers an opportunity for inventors to have Staples manufacture their ideas and pay them royalties.

One recent winner, Sarah Pantaleo of Harleysville, Pennsylvania, submitted an idea for a compact-disc storage system that Staples now markets under the brand name Spindex. In 2004, inventor Eric Gibbons won the contest with a product called WordLock. This is a new type of combination lock, for such things as bicycles and luggage, that uses easy-to-remember words instead of numbers. As well as winning the Staples contest, WordLock was voted a top-100 invention in the Invent Now America competition.

PRESTIGIOUS CONTESTS

Some contests are prestigious affairs designed to recognize the world's greatest inventors. Sponsored by the Lemelson Foundation and administered by the Massachusetts Institute of Technology (MIT), the Lemelson-MIT Awards are sometimes known as the Oscars for inventors. The top award, the Lemelson-MIT Prize, is worth $500,000 and is one of the world's biggest single awards for established inventors.

The craft SpaceShipOne *is tethered to the bottom of its launch ship,* WhiteKnight, *as it enters suborbital space in September 2004.*

Past winners of the Lemelson-MIT Prize include Dean Kamen (1951–), whose inventions include the Segway personal transporter and iBOT wheelchair; and Douglas Engelbart (1925–), a pioneer of personal computing. Between 1995 and 2006, the Lemelson-MIT program also awarded a prize for outstanding lifetime contributions to inventing. It has been won by Raymond Damadian (1936–), the inventor of magnetic resonance imaging (MRI), a medical scanning technology, and Stephanie Kwolek (1923–), developer of Kevlar, a tough synthetic material. Recognizing the growing importance of environmental issues, the Lifetime Achievement Award has now been replaced by a new $100,000 Award for Sustainability.

The X PRIZE is another prestigious award. The first contest challenged inventors to build a reusable space plane and to fly into space and back twice within two weeks. Several teams competed in the autumn of 2004 before the prize was won that October by an experimental craft, *SpaceShipOne*. The X PRIZE Foundation plans to hold further contests to encourage technological breakthroughs in automobile design, medicine, education, and environmental protection.

ENCOURAGING PROMISE

Whereas the Lemelson-MIT Prize rewards proven inventions, other contests are designed to identify and nurture upcoming ones. Hammacher Schlemmer, a product-testing organization, runs the annual Search for Invention contest which is open only to patented products that are not yet being manufactured. The contest has four categories: recreation, personal care, electronics, and the home and garden. Like many other contests, Search for Invention recognizes that invention is about much more than having good ideas. Thus, prizes are awarded not just for originality, but also for turning ideas into winning designs, market feasibility, safety, and overall benefit to users.

Many contests are aimed specifically at encouraging young people. Apart from the $500,000 main prize, the Lemelson-MIT Awards also include smaller prizes for young inventors. Each year, the $30,000 Student Prize goes to a college senior or graduate student who has developed something that demonstrates "remarkable inventiveness." Carl Dietrich, a doctoral student at MIT, won the 2006 award for a series of promising inventions, including a new type of flying car and a lower-cost rocket engine. The previous year, a former MIT student, David Berry, won the prize for developing innovative treatments for stroke and cancer patients.

Younger children also have the opportunity to compete in inventing contests run by such companies as Coca-Cola, LEGO, and Xerox. The FIRST LEGO

A contestant in the FIRST LEGO competition works on his robot.

League is a national contest that challenges children ages 9–14 to design toy robots that can solve serious, real-world problems. In the 2005 contest, which focused on how little is known about the oceans, teams had to build robotic devices that could study the seabed. The annual Team America Rocketry Challenge offers similar challenges. Instead of sea-sweeping robots, middle- and high-school students have to make a rocket that can carry a raw egg 850 feet into the air and bring it back down unbroken.

UNUSUAL CONTESTS

Unlike the Lemelson-MIT and Hammacher Schlemmer awards, which accept a broad range of entries, other contests are more narrowly defined. The DuPont Chemical Company offers the Plunkett Awards, named for Teflon's inventor, Roy Plunkett (1910–1994), which is designed to recognize new fabrics made using DuPont chemicals. Another contest, the Craftsman/NSTA Young Inventors Award Program, chal-

lenges students to devise a new or improved tool. National Engineers Week each year challenges children to design a city of the future. Some contests are even more specific. A 2004 study by the BoatU.S. Foundation found that many people fail to wear a life jacket because they find it too uncomfortable. This prompted the organization to launch its Innovations in Life Jacket Design Competition, which encourages inventors to design a life jacket that people will actually enjoy wearing.

While most contests have a serious purpose, some encourage people to take the inventing process much more lightheartedly. The annual Rube Goldberg Machine Contest honors cartoonist Reuben Goldberg (1883–1970), who was famous for making amusing cartoons of improbable contraptions. It presents a simple problem, such as putting toothpaste on a toothbrush, shutting off an alarm clock, or making a cup of coffee. The catch is that they must solve the problem in the most absurd and elaborate way possible—typically involving at least twenty different stages. One

Invention Contests

Contest	Web site	Entry Specifications	Prize
ExploraVision	http://www.exploravision.org/	This contest is open to teams of students in grades K to 12. Students simulate a research and development team to create technologies they think will be useful in everyday life.	Four teams at the national level win the first prize, a $10,000 U. S. savings bond.
FIRST LEGO League	http://www.firstlegoleague.org/	This contest is open to students from ages 9 to 14. The students must use Legos to build a robot according to the organizers' rules.	Prizes are awarded in four categories: technical, team presentation, special recognition, and judges' awards.
FIRST Robotics Competition	http://www.usfirst.org/robotics/	This robot-building contest is an international competition for high school students and awards prizes based on various criteria.	Awards are given in various categories.
Future City Competition	http://www.futurecity.org/	A contest for 7th and 8th graders to design and build a city of the future.	Regional champions attend the national competition in Washington, D. C., where they compete for the grand prize of a trip to U. S. Space Camp in Huntsville, Alabama.
Intel International Science and Engineering Fair	http://www.sciserv.org/isef/	This science fair is open to students in grades 9 to 12. Students can either invent a new technology or present a study about a particular subject.	Various scholarships are awarded.
Invention Quest	http://inventionquest.dja.com/	This contest to design products that will simplify everyday life is open to inventors of all ages. There are different contests for different ages.	The winning inventor in each age group receives $25,000.
Invent Now America	http://www.inventnow.org	This contest encourages entrants to create new items that are considered by the judges of the contest to be technological or scientific innovations. It is open to everyone.	The grand prize is $25,000.
InvenTeams	http://web.mit.edu/inventeams/	This is a program in which high school students can apply for grants of up to $10,000 to identify and research a problem and then develop a prototype invention as a solution.	There are no prizes awarded.
Rube Goldberg Machine Contest	http://news.uns.purdue.edu/ UNS/rube/rube.index.html	Entrants must complete a very simple task using the most absurd and complicated solution they can think of and in no less than twenty steps. Open to high school and college students.	There are two prizes for the contest: one is a trophy, awarded by a panel of judges; and the other is the People's Choice Award.
Team America Rocketry Challenge	http://www.aia-aerospace.org/ aianews/features/ team_america/	A model rocket contest for students enrolled in grades 7 to 12. The goal is to build a model rocket that carries one raw egg to an altitude as close to 850 feet as possible and stays airborne for as close to 45 seconds as possible.	Teams with the top 100 scores will compete for a share of the $60,000 prize package in a national fly-off.
ThinkQuest	http://www.thinkquest.org/	Students ages 9 to 19 build innovative Web sites on any educational topic.	The prize is a cash award for their school and a new laptop computer for student participants.
Siemens Competition in Math, Science, and Technology	http://www.siemens-foundation .org/competition/	High school students competing in this contest research a variety of projects either as individuals or as part of a team.	A number of scholarships are awarded for the winning entries.
Wild Planet	http://www.kidinventorchallenge .com/	This toy-inventing contest is for children ages 6 to 12.	One hundred children are chosen as winners and get to be "kid consultants" for Wild Planet Toys for a year. In addition to this, they will get the opportunity to develop their winning idea into a toy.

recent challenge, to invent a device that can pour a cup of water, produced all kinds of inspired entries, including a machine that combined a toaster, smoke detector, microphone, and tennis racket. Other contests try to tap into people's creativity

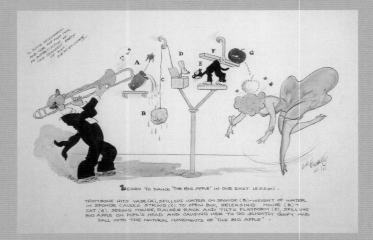

In this 1937 Rube Goldberg cartoon, a woman learns a dance called the Big Apple: "Trombone hits vase (A), spilling water on sponge (B)—weight of water in sponge causes string (C) to open box, releasing mouse (D)— cat (E), seeing mouse, raises back and tilts platform (F), spilling big apple on pupil's head and causing her to go slightly goofy and fall into the natural movements of 'The Big Apple.'"

in a similar way. The Soriano-Dietz Invention Contest recently challenged inventors to come up with a new gadget made entirely from paper clips. The inspiration for this unusual competition is a well-known test of creativity, where people are challenged to come up with as many uses for a paper clip as they possibly can.

Whether serious or frivolous, contests help to remind people that inventors play a central part in solving humankind's problems. The more prestigious contests, such as Lemelson-MIT, recognize distinguished and promising inventors whose contributions sometimes go unnoticed. On the other hand, contests like the Rube Goldberg help to remind people that science and technology—in the shape of everyday inventions—can be both useful and fun.

—Chris Woodford

Further Reading

Books

Levy, Richard. *The Complete Idiot's Guide to Cashing In on Your Inventions*. Indianapolis, IN: Alpha, 2001.

Sobey, Ed. *How to Enter and Win an Invention Contest*. Hillside, NJ: Enslow, 1999.

Von Oeck, Roger. *A Whack on the Side of the Head*. New York: Warner Business Books, 1998.

Web sites

American Inventor
> Web site of the ABC Television inventing contest.
> http://abc.go.com/primetime/americaninventor/

Invent Help
> Practical advice for inventors, including lists of inventing contests.
> http://www.inventhelp.com/

Kid Inventor Challenge
> An annual contest run by Wild Planet Toys.
> http://www.kidinventorchallenge.com/

Lemelson-MIT Prize
> Details of the "Oscar for inventors," with lists of past winners.
> http://web.mit.edu/INVENT/a-prize.html

Rube Goldberg
> Includes examples of Rube Goldberg's cartoons and details of the Rube Goldberg Machine Contest.
> http://www.rube-goldberg.com/

See also: Appert, Nicolas-François; Damadian, Raymond; Engelbart, Douglas; Harrison, John; History of Invention; Kamen, Dean; Kwolek, Stephanie; Patents; Plunkett, Roy J.; Young Inventors.

MARTIN COOPER

Developer of the cell phone

1928–

The telephone and radio were two of the greatest inventions of the late 19th century. During the 20th century, they came together in a third great invention: the cellular (mobile) telephone. Cellular phone use has grown rapidly since the device was developed in the 1970s; roughly one billion mobile handsets are now in use worldwide. Although many individuals contributed to the development of cell phones, American inventor Martin Cooper is usually credited with making the handheld cell phone a reality.

EARLY YEARS

When Martin Cooper was born on December 26, 1928, the modern electronic tools that we now take for granted—personal computers, color televisions, pocket calculators, and compact-disc players—had yet to be invented. "It's really embarrassing," he recalled later, "The television did not become commercial until I was past my teen years." Growing up in Chicago, Cooper—like many children did in those days—entertained himself by reading books, especially science fiction.

At that time the most exciting new technology was radio. People called it "wireless" and thought of it mostly in connection with radio programs and, later, with television (because radio waves are also used to transmit the images seen on a television screen). Few people had any idea that radio had other uses. Cooper believed, however, that radio was the future. After earning a degree in electrical engineering at the Illinois Institute of Technology, he served in the navy during World War II, and then joined the research department of a telecommunications company.

RADIOS AND TELEPHONES

The road to cellular phones began with Alexander Graham Bell's invention of the telephone in 1876. Around the same time, scientists were

Martin Cooper poses with the first cell phone in 2003.

discovering how radio waves could send information through the air without using wires. The first person to transmit human speech by radio waves was American engineer Reginald Fessenden (1866–1932). On December 24, 1906, Fessenden used radio to send a spoken message 11 miles from Brant Rock, Massachusetts, to ships in the Atlantic Ocean.

Personal Communication

When Alexander Graham Bell invented the telephone in 1876, he was trying to solve a specific problem: how to carry speech down a wire using electricity. The arrival of fax machines, and later the Internet, made sending all kinds of information (or data) along telephone wires possible. Once simply a way of carrying voices, the phone network is increasingly seen as a "multimedia pipeline" along which almost limitless information can travel.

A similar transformation has happened with cell phones. When Martin Cooper developed his first cell phone in the 1970s, it was designed for people like taxi drivers who needed to make calls on the move. The arrival of digital cell phone technology in the early 1980s made it possible to send and receive any kind of digital (numerically coded) data between two cell phones or between a cell phone and another computer-based gadget. Thus, many cell phones now have built-in digital cameras that can send and receive photos and video clips. Most can send e-mail and browse Web pages. Some have built-in MP3 players, video players, digital radios, and even tiny digital televisions. Cell phones started out as mobile telephones; they have rapidly become go-anywhere, do-anything personal communicators, firing streams of digital data back and forth through the air.

At a fraction of the size of Cooper's original phone, modern cell phones are also far more powerful and are able to serve as cameras, music players, and personal digital assistants (PDAs).

The first truly mobile phones, called radiotelephones, appeared by the 1940s; they were cumbersome radio sets carried around in emergency vehicles and taxis. In 1946, AT&T and Southwestern Bell introduced MTS (Mobile Telephone System), a commercial radio system that used five antennas and a central switching office to transmit conversations between moving vehicles. The following year, AT&T introduced long-distance radiotelephone service between New York City and Boston.

These early systems were so popular that they had waiting lists of up to 10 years to buy them. They were limited in that they could transmit only a few telephone calls simultaneously. Early radiotelephones sent and received calls on a narrow section of radio-wave frequencies known as a band. This was like a pipe of limited diameter through which only a certain number of calls could squeeze at a time.

INVENTING THE HANDHELD CELL PHONE

Clearly many people wanted radiotelephones; in response, the Federal Communications Commission (a U.S. government body that regulates telecommunications) agreed to make more frequencies available if telephone companies could find a more efficient way to use them. Two companies, AT&T Bell Labs and Motorola, took up the challenge. In the late 1960s, Bell Labs developed a system (Metroliner) that enabled public pay telephones on trains to send and receive calls by radio. Motorola, where Martin Cooper had been working since 1954, was developing a completely mobile phone.

At Motorola, Cooper had led a team of scientists who developed walkie-talkie radios, used for the first time by the Chicago police department in 1967. Cooper's walkie-talkies were comparatively simple, because only a limited number of radios were in use at once. Developing a full-scale radio telephone system was trickier. What was needed was a way of allowing many different telephones to use the same band of radio frequencies but to keep the different calls from interfering with each other.

Cooper solved this problem by using a system called "cells," which had been invented by Bell Labs in the 1960s. The idea was to place radio

TIME LINE

1928	1954	1967	1973	1975	2006
Martin Cooper born in Chicago.	Cooper joins research department at Motorola.	Cooper's team develops walkie-talkie radios.	Cooper's team develops the DynaTAC.	Cooper's team at Motorola awarded a patent for a "radio telephone system."	Cooper serves as chairman of ArrayCom.

The first page of Cooper's patent for his radio-telephone system.

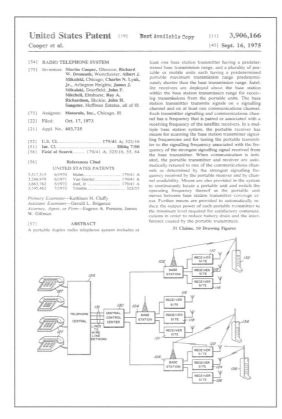

antennas all around a city in a honeycomb pattern. Each antenna would serve only the small area, or cell, immediately around it, not the entire city. Neighboring cells could use the same radio frequencies, so as the number of antennas and cells increased, the number of telephones that could be used at the same time would increase also.

Motorola's first cell phone was the shape and size of a brick and weighed 28 pounds (12.7 kg). However, Cooper's team soon took advantage of new microelectronic technology to build a smaller device, the DynaTAC, which weighed only 2.8 pounds (1.27 kg). On April 3, 1973, Cooper walked through the streets of New York City to demonstrate the new phone. He decided to test it one last time and, on the spur of the moment, called his rival at Bell Labs, an engineer named Joel Engel, to boast of his achievement. Of this, the world's first cell phone conversation, Cooper said later, "I don't remember my precise words. I do remember a kind of embarrassed silence at the other end." The patent for a "radio telephone system" was granted to Martin Cooper and seven of his Motorola colleagues on September 16, 1975. Telephones had caused a revolution at the end of the 19th century; mobile cell phones caused a similar stir some hundred years later. By the early 1980s, Cooper's team at Motorola was running a large-scale test service in Washington, D.C., and Baltimore, Maryland. DynaTAC, the world's first commercial handheld cellular phone, finally went on sale in 1984. This first-generation system weighed 28 ounces (793 g) and cost $3,995, and the battery had a one-hour life.

MOBILE REVOLUTION

Technology advanced greatly during the 1980s. In Europe, 26 telephone manufacturers got together and launched a joint cell phone system known as GSM (Global System for Mobile Telecommunications). It used digital technology: calls were transmitted not as raw sounds but as coded strings of numbers. This enabled many more telephone calls to be handled. It offered other advantages, too: digital phone calls had better sound quality, had greater protection from eavesdropping, and could send and receive short written messages (known as SMS or text messages). In the United States, the PCS (Personal Communications Services) system worked in a similar way. These systems came to be known as second-generation networks.

In the 21st century, mobile networks are faster, the handsets are smaller, and all kinds of information (from photos to Web pages) can be sent to and from a cell phone. Many developing countries are abandoning landlines entirely; residents use wireless cell phone networks for both their telephone and their Internet access.

Several decades after making the world's first cell phone call, Martin Cooper is still too busy to retire: "Think about what the alternative is," he told one reporter. "You could sit around talking about the past and boring people to death, or you can keep active and be where the action is." In 2006 he was the chairman of his fifth start-up company, ArrayCom, which develops advanced cell phone networks.

—Chris Woodford

Further Reading

Books

Agar, Jon. *Constant Touch: A Global History of the Mobile Phone*. New York: Totem, 2005.
Bridgman, Roger. *Eyewitness: Electronics*. New York: Dorling Kindersley, 2000.
Levinson, Paul. *Cellphone: The Story of the World's Most Mobile Medium and How It Has Transformed Everything!* New York: Macmillan, 2004.

Web sites

How Cell Phones Work
    An introduction to cell phone technology.
    http://electronics.howstuffworks.com/cell-phone.htm
Privateline.com: Digital wireless
    An introduction to cell phones and their history.
    http://www.privateline.com/PCS/splash.htm

See also: Bell, Alexander Graham; Caselli, Giovanni; Communications; Marconi, Guglielmo.

ALLAN CORMACK AND GODFREY HOUNSFIELD

Inventors of the CT scanner

1924–1998 and 1919–2004

In 1979 the Nobel Prize committee wanted to honor the inventor of the computed tomography scanner, or CT scanner, which revolutionized medicine by allowing doctors to take extremely detailed images of the insides of patients. The problem was that at least four people could be said to have discovered the principles behind the CT scanner. The committee finally chose to honor two men who had never worked together and had never even met: Allan Cormack and Godfrey Hounsfield. Although Cormack was the first to realize how a CT scanner could work, Hounsfield was the first to build a working scanner. Ever since the joint prize was awarded, the two men have been linked in medical history.

EARLY YEARS: A PHYSICIST IN SOUTH AFRICA

Allan MacLeod Cormack was born in South Africa to Scots parents in 1924. Cormack's father was an engineer with the post office, and the family moved often as his father was assigned to different locations. Cormack's father died in 1936, and the family then settled in Cape Town.

As a teenager, Cormack was fascinated by astronomy, but once he entered the University of Cape Town he decided it would be difficult to make a living as an astronomer, and he chose to study engineering instead. As his schooling progressed, Cormack discovered that he preferred physics, and he earned a bachelor's and then a master's degree in that discipline.

He moved to Cambridge University in England briefly for postgraduate study; there, he met his future wife. Cormack felt he had to have a job that offered a salary high enough to support a family before he could marry, so he contacted the physics department at the University of Cape Town, which offered him a teaching position. Cormack, now married, returned to South Africa in 1950.

A CALL FROM A HOSPITAL

While working in Cape Town in 1955, Cormack received a call from the local Groote Schuur Hospital. At the time, South African law required that hospitals employ a physicist to supervise the use of radioactive isotopes in radiation therapy for patients suffering from diseases such as cancer. Groote Schuur's physicist had just resigned, and the hospital had called to see if Cormack would take the part-time job in addition to his teaching. He accepted.

During the six months he worked at Groote Schuur, Cormack noticed a rather significant shortcoming of radiation therapy. When attempt-

Allan Cormack in 1979.

ing to determine how large a dose of radiation to give a patient, doctors relied on charts that were based on the assumption that every kind of tissue in the human body—bone, muscle, skin, fat—absorbed radiation equally. This was an assumption that Cormack and the doctors knew was false—materials of different density absorb radiation differently.

Cormack decided to create what would be essentially a map of the body that showed exactly how much radiation each kind of tissue absorbed. Patients would then receive no more radiation than they actually needed to treat their conditions.

BREAKTHROUGH AND FRUSTRATION

While trying to figure out how to create this map, Cormack set up an experiment in 1957 in which he shot a beam of gamma rays through an aluminum-and-wood cylinder at a variety of angles into a detector. The experiment produced unusual data, and when Cormack investigated, he discovered that the aluminum cylinder, which he had thought was of a uniform thickness, in fact had a core peg of a density slightly different from the rest of the cylinder. Cormack had unwittingly detected this difference with his experiment.

Cormack quickly realized that a scanner that could detect such minute differences in density could have many important medical uses, particularly for diseases of the brain. One major drawback of traditional x-rays is that they cannot show the brain of a living patient, because the dense bone of the skull blocks the radiation (see box, How a CT Scanner Works). Before the advent of the CT scanner, if a person was suspected of having a brain tumor, physicians would have to perform dangerous exploratory surgery before they knew exactly what was wrong.

Cormack set up another experiment, this time crafting a cylinder with an aluminum outer ring, a Lucite interior, and two aluminum pegs. Aluminum is denser than Lucite, and the cylinder was intended to represent the skull, the brain, and two tumors, respectively. By shooting radiation at the "skull" from many different angles, he easily found the "tumors" within. Cormack, who had by now taken a position as a physics professor at Tufts University in Medford, Massachusetts,

How a CT Scanner Works

A CT scanner and an x-ray machine obtain information by shooting radiation through a patient's body; however, a CT scanner is able to create far more detailed pictures of a patient's organs.

A conventional x-ray machine takes a picture as a camera would: an x-ray emitter is placed on one side of the body part of interest, a piece of film is placed on the other side, and a photograph-like image is created. That process, however, limits the quality of the image in two significant ways. First, x-ray film is not an especially sensitive detector of x-rays. Second, the resulting image is a composite of several images created as the x-rays pass through layers of organ and bone. Therefore, denser objects, such as the bones of the skull, conceal objects that are less dense, such as the brain.

In contrast, a CT scanner does not take a single reading of the area of the body being investigated; as the emitter is revolved around the body, extremely sensitive detectors take readings from many different angles. The result is an enormous quantity of data; a CT scanner typically produces tens of thousands of readings. To then create an image, the scanner requires a powerful computer to run the many sophisticated algorithms needed to translate all those readings into a single image.

Once the image is complete, it is remarkably detailed. It is almost impossible to tell one organ from another on an x-ray; but on a CT scan, different organs are easy to discern. Whereas a conventional x-ray of a head will simply show a skull, a CT scan will show the skull and the brain in tremendous detail. In addition, the digital image produced by a CT scan can be manipulated on a computer, so that a three-dimensional image from a CT scan can be virtually rotated or sliced as needed.

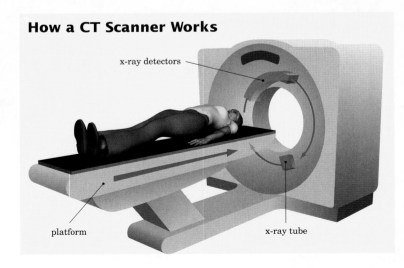

How a CT Scanner Works

x-ray detectors

platform

x-ray tube

CT scans allow doctors to form three-dimensional images of a part of a patient's body.

published two articles in the journal *Applied Physics* on his ground-breaking discovery, one in 1963 and one in 1964.

"There was virtually no response," he later recalled. The implications of the physicist's experiment seemed to be unappreciated by manufacturers of medical devices. Part of the problem was that the technology was clearly going to require expensive, powerful computers to work, and no manufacturer was willing to invest in an untried technology. In addition, Cormack was busy with his job; named chairman of the physics department at Tufts in 1968, he did not have the time to vigorously promote what was essentially a side project.

Cormack's breakthrough seemed fated to remain obscure. He was not alone: at least two other scientists have also been credited with discovering many of the underlying principles of the CT scanner, but they, too, were unable to interest any company in manufacturing the new technology. Only when Hounsfield made a similar breakthrough did this situation change.

EARLY YEARS: A COUNTRY CHILDHOOD

Born in 1919, Godfrey Newbold Hounsfield grew up on a farm near Nottinghamshire, England. The youngest of five children, Hounsfield showed an early interest in machinery and science: as a teenager he built a record player out of spare parts, and from time to time he jumped off the tops of haystacks wearing a homemade glider. He was less interested in his schoolwork, however, and he did not do well enough to be admitted to a university.

When World War II broke out, Hounsfield's interest in airplanes led him to volunteer for the Royal Air Force (RAF). He started taking RAF courses in radio and radar mechanics, and he showed such aptitude that he was made a lecturer on radar. Following the war, Hounsfield received a grant that allowed him to attend Faraday House Electrical Engineering College in London, from which he graduated in 1951.

Godfrey Hounsfield, photographed in 1979.

TIME LINE

1919	1924	1939–1945	1950	1951	1957	1958
Godfrey Newbold Hounsfield born near Nottinghamshire, England.	Allan MacLeod Cormack born in South Africa.	Hounsfield lectures on radar in Royal Air Force during World War II.	Cormack accepts teaching position at University of Cape Town.	Hounsfield graduates from engineering college and begins work at EMI.	Cormack begins imaging experiments.	Hounsfield's team at EMI builds Great Britain's first all-transistor computer.

THE RIGHT MAN IN THE RIGHT PLACE

That same year, Hounsfield went to work for EMI, a London music company that had branched out into electronics. Beginning around 1958, Hounsfield led a team at EMI that built Britain's first all-transistor computer. A few years later, Hounsfield was given the opportunity to pick a field of research, and he chose pattern recognition.

In 1967, while thinking about various problems in pattern recognition, Hounsfield made a mental leap similar to the one Cormack had made. Hounsfield realized that penetrating an object with radiation at different angles would produce data that could be used to construct an image of the object's interior.

Unlike Cormack, Hounsfield did not spend the next several years vainly trying to interest people in his project. As a respected researcher at a technologically oriented company that was willing to fund his ideas, Hounsfield had the means to build and perfect the first CT scanner.

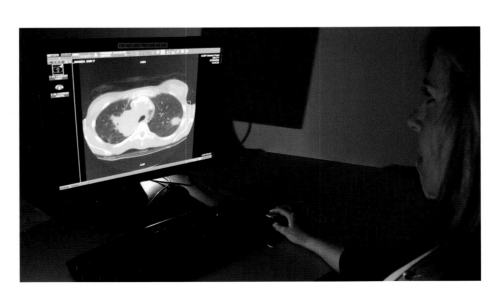

In 2004 a radiology specialist reviews a CT scan of a cancerous lung tumor at the Comprehensive Cancer Center at the University of California–San Francisco.

TIME LINE

1963–1964	1967	1968	1972	1979	1998	2004
Cormack publishes papers on his experiments.	Hounsfield gets idea for CT scanner and begins experiments.	Cormack named chairman of Tufts University's physics department.	EMI unveils the first commercially available CT scanner.	Cormack and Hounsfield awarded the Nobel Prize.	Cormack dies.	Hounsfield dies.

Hounsfield began much like Cormack, shooting gamma rays at inanimate objects. Soon he switched to passing x-rays through the brains of humans and cows. The clarity of the resulting images was startling: the scans showed the interior structure of the brain in remarkable detail. Hounsfield then conducted his first scan on a human patient, a woman who was suspected of having a brain tumor. Although the scan took 15 hours, the results were impressive—the tumor showed up as a large, dark spot that was easy even for someone with no medical training to see. The woman had surgery to remove the tumor.

A REVOLUTION IN MEDICAL IMAGING

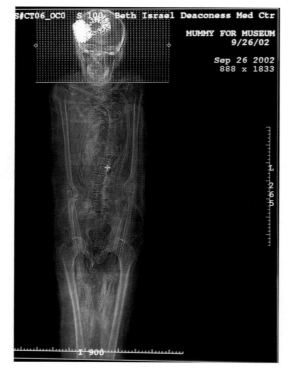

The CT scan of an unidentified mummy from the Michael C. Carlos Museum at Emory College. CT scans can be used to determine the age, gender, diet, and overall health of mummies.

In 1972 EMI unveiled the first commercially available CT scanner to the world. Despite its high price, initially more than three hundred thousand dollars, the device caused a sensation. Soon other medical-device firms were offering their own scanners and looking to previously neglected scientists like Cormack for help in improving the devices.

Cormack and Hounsfield met in Sweden seven years later, when they received the 1979 Nobel Prize in Physiology or Medicine. Although both men spoke out in favor of advancing medical imaging, neither one radically altered his career to pursue the field. Cormack remained a professor at Tufts and died in 1998; Hounsfield remained at EMI, was knighted in 1981, and died in 2004.

In the late 1980s two new scanners were invented that expanded the functionality of the CT scanner: a superfast scanner that can take pictures of a heart between beats, and a spiral CT scanner that can capture images of large portions of the body in a very short time.

Thousands of CT scanners are now in use in the United States, and are increasingly applied for nonmedical purposes. Archaeologists use CT scanners to examine the interior of historical artifacts; security personnel at airports and elsewhere use them to scan suspect packages and bags. Indeed, CT scans have become so popular in recent years that medical organizations have felt the need to suggest limits on their use because of worries that patients will be overexposed to x-rays.

These scanners have affected many medical fields, but none more than neurology, which specializes in diseases of the brain. From 1979 to 1995, the number of deaths from brain disease in the United States plummeted 37 percent, largely because of the improvements in imaging technology that were separately envisioned by Cormack and Hounsfield.

—Mary Sisson

Further Reading

Books

Kevles, Bettyann Holtzmann. *Naked to the Bone: Medical Imaging in the Twentieth Century.* New Brunswick, NJ: Rutgers University Press, 1997.

Wolbarst, Anthony B. *Looking Within: How X-ray, CT, MRI, Ultrasound, and Other Medical Images Are Created, and How They Help Physicians Save Lives.* Berkeley: University of California Press, 1999.

Web sites

Computed Tomography (CT or CAT Scan)
    Descriptions of a variety of modern CT procedures by the Radiological Society of North America.
    www.radiologyinfo.org/sitemap/modal-alias.cfm?modal=CT
CT Scan
    Mayo Clinic Web page showing CT scans of various organs.
    www.mayoclinic.com/health/ct-scan/FL00065

See also: Damadian, Raymond; Health and Medicine.

CORPORATE INVENTION

One of the world's greatest inventors, Thomas Edison (1847–1931), filed an astonishing 1,093 patents during his 84 years. By contrast, one of the world's most inventive companies, the IBM corporation, filed 2,941 patents in 2005 alone. These figures demonstrate the reality of modern inventing: individual inventors have largely been replaced by creative corporations. Although some inventions are still being credited to individuals, most modern inventions are developed by teams of people in large corporations, often with no single person responsible for their success.

THE BIRTH OF COPORATE INVENTING

The idea of research laboratories originated in Germany at the end of the 19th century, pioneered by chemical companies such as Bayer, BASF, and Hoechst. However, Thomas Edison popularized the concept of a group approach to invention. In Edison's eyes, a research laboratory was an invention factory. However, the people he employed in his world-famous laboratories at Menlo Park and West Orange in New Jersey were largely there to help Edison develop his own remarkable inventions, including the electric lightbulb, the phonograph (a music player), and the movie camera. His employees were not expected or encouraged to work on their own inventions.

The industrialist and prolific inventor Henry Ford (1863–1947) began his working career as one of Edison's employees and later became one of his friends. Like Edison, Ford understood that research was an investment in the future. As a tribute to his friend, he re-created Edison's Menlo Park laboratory as an exhibit in his museum at Dearborn, Michigan, in the 1920s. Ford also built a research laboratory of his own in 1929. There Ford's engineers developed a new type of plastic based on

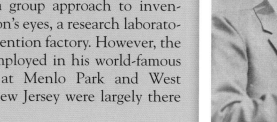

Henry Ford, left, and his friend Thomas Edison, photographed in 1921.

soybeans, fashioning an experimental car body from this new material in 1941. Although the idea was a success, it was abandoned during World War II when the company devoted itself to the war effort.

During the first half of the 20th century, many other corporations established their own laboratories. In 1900, General Electric (formed in a merger between Edison's company and its main rival) set

Bell Labs

During the 20th century, no corporate research laboratory anywhere in the world could match Bell Laboratories ("Bell Labs") for its dazzling record of innovation. Since its creation in 1925, Bell Labs has developed no fewer than forty thousand inventions. According to a Bell Labs estimate, a typical home contains "at least 25 products" based on technologies invented by its scientists, including digital cell phones, computers, CD players, and transistor radios.

As its name suggests, Bell Labs evolved from the pioneering work of telephone inventor Alexander Graham Bell (1847–1922). Beginning with 3,500 employees, who occupied an imposing building on West Street in Manhattan, Bell Labs later relocated to a huge complex in Murray Hill, New Jersey.

Bell Labs was innovative from the start. In 1925, it made the first public demonstration of the modern, telephone-based facsimile (fax) machine, an idea pioneered about seventy years earlier by Italian science teacher Giovanni Caselli (1815–1891). By the 1940s, it developed the first mobile radiotelephones. In 1947, Bell's most famous invention, the transistor, was created. The following year, a Bell scientist, Claude Shannon (1916–2001), outlined his "information theory," which delineated the most efficient way of sending data between, for example, computers and telephones. Later Bell inventions included lasers, developed by Arthur Schawlow (1921–1999) and Charles Townes (1915–), light-emitting diode (LED) displays, the UNIX computer operating system, digital cell phones, and the C and C++ computer programming languages.

In the 1990s, AT&T turned Bell Laboratories into a separate company, Lucent Technologies. The company continues to develop communication and computer technologies such as fiber optics, lasers, and better forms of transistors.

up a major research laboratory at Schenectady, New York. In 1911 AT&T created what could be called the world's most famous research center, Bell Laboratories (see box, Bell Labs). Such places were typically staffed with brilliant academics hired away from universities.

Inspired by the success of its German rivals, the American corporation E. I. du Pont de Nemours & Co. (generally called DuPont) built an impressive research center in Wilmington, Delaware. It began by attracting German chemists to the United States with salaries 10 to 15 times greater than those they were earning at home. DuPont also focused its efforts on employing promising young academics at the beginnings of their careers. Wallace Carothers (1896–1937) joined DuPont from Harvard University. In Wilmington, Carothers led a team of other scientists investigating the properties of plastics. With increasing pressure from DuPont's management to produce commercial products, Carothers and his team shifted from pure science to applied technology—with spectacular success. Their greatest triumph was the invention of nylon, a family of revolutionary plastics used to make textiles in the 1930s. In 1938, another successful material, trademarked as Teflon, was developed by DuPont's chemist Roy Plunkett (1911–1994).

LEARNING FROM THE WAR

During World War II, thousands of scientists and engineers (ten thousand in the United States alone) redeployed their attention to military research. Some went to work in places such as the Naval Research Laboratory, a military invention factory set up in 1923 at Edison's suggestion. Bringing creative scientists and engineers together to develop military technology proved crucial to the Allies ultimate victory, spurring the development of atomic energy and computers—inventions that would also have important peacetime applications.

After the war, the scientists returned to their research. Many were eager to resume university courses or careers interrupted by the war; others moved into industrial research and became professional inventors. For many scientists, their experiences during the war became

The Naval Research Laboratory supports high-tech research with military applications, such as this waterproof laptop, unveiled in 2003.

part of their scientific training and led many of them into rewarding careers in industry.

MODERN TIMES

In the postwar United States, corporate inventing became standard for many research laboratories. In 1947, three physicists who had worked in wartime naval research, John Bardeen (1908–1991), Walter Brattain (1902–1987), and William Shockley (1910–1989), invented the transistor—a revolutionary new kind of electronic amplifier and switch—for Bell Laboratories. About a decade later, Jack Kilby (1923–2005) developed the integrated circuit, an improved way of making transistors, for Texas Instruments.

Most of the world's greatest inventions have since been conceived and brought to fruition by corporate research laboratories. Following its success with

nylon, DuPont produced many other pioneering materials after World War II, including a strong and durable textile called Kevlar. The incredibly stiff fibers used to make Kevlar were discovered in the 1960s by Stephanie Kwolek (1923–) and further developed by a team of DuPont researchers working with her. The 3M Corporation has been similarly innovative, with notable products including Scotchgard and Post-it notes. Both products are widely recognized as 3M inventions, with the names of their inventors (Patsy Sherman, Scotchgard; Spence Silver and Arthur Fry, Post-its) largely unknown. The Motorola Company is credited with making the first mobile cellular phone in 1973, though the patent for that invention carries the names of a team of Motorola engineers, led by Martin Cooper (1928–). Another great invention in communications technology, the fiber-optic cable, was inspired by the scientific research of Indian physicist

At the 3M Innovation Centre in Singapore, a researcher demonstrates a new product to George Yeo, Singapore's former minister for trade and industry; and Livio DeSimone, the former chairman of 3M.

Narinder Kapany (1927–) in the 1950s. However, the Corning Glass Company is generally credited with turning fiber optics into one of the world's most important modern technologies almost twenty years later.

IDEAS THAT BUILD CORPORATIONS

Some of the most successful companies of modern times have grown out of great inventions. During the 19th century, John Deere (1804–1886) and Cyrus Hall McCormick (1809–1884) developed important agricultural machines and built companies to profit from their inventions. George Eastman (1854–1932) began his photographic empire in 1892 to market his easy-to-use Kodak cameras. At the start of the 20th century, both Thomas Edison and Henry Ford were developing huge corporations on the strength of superior ideas, a trend that has continued ever since.

High-technology electronics and computer companies are good examples of this trend. William Shockley set up his own company to capitalize on the invention of the transistor in the 1950s. His venture was a commercial flop. Nevertheless, Shockley's employees included a few first-rate scientists who went on to establish some of the world's greatest electronics companies, including Intel. Shockley is generally credited with helping to start Silicon Valley, the area in California where many electronics companies are based; consequently, he is recognized as a pioneer of corporate invention.

Yahoo! founders David Filo (left) and Jerry Yang (right) at the company headquarters in Santa Clara, California.

More recently, Silicon Valley has witnessed the birth of Apple Computer, eBay, Google, and Yahoo!—all built on great ideas. Apple was formed in 1976 when two computer hobbyists, Steve Jobs (1955–) and Steve Wozniak (1950–), started making easy-to-use home computers in a garage belonging to Jobs's parents. Yahoo! was founded in 1994 after Jerry Yang (1968–) and David Filo (1966–), both Stanford postgraduates, began compiling a popular Web site, "Jerry and David's Guide to the World Wide Web." By the late 1990s, their Web site had evolved into one of the world's biggest media corporations. Google began in a similar way, when two other Stanford postgraduates, Larry Page (1973–) and Sergey Brin (1973–), had an idea for making a more effective Internet search engine. Like Yahoo! their invention rapidly became a multibillion dollar corporation. eBay is another Silicon Valley success story. The popular Internet auction Web site was born from a simple idea: programmer Pierre Omidyar (1967–) realized that the Web would make a great virtual marketplace to exchange goods.

INVENTORS AND BUSINESSPEOPLE

Those who conceive great inventions are not always the best people to capitalize on them. Scientists and engineers who excel at turning scientific discoveries into practical technologies may not have the skills to run a large corporation.

Some inventors, such as Robert Noyce (1927–1990), one of the founders of Intel, discover they have a talent for running their own corporations. Others, such as William Shockley, Noyce's original employer, find they are completely unsuited to the business world. Still other inventors succeed initially in setting up a company, only to fall into difficulties when competitors catch on to the invention. Karl Benz (1844–1929), one of the inventors of the automobile, decided to resign as the head of his own firm after rivals began to challenge his company.

Problems can also occur after a successful company goes public, selling shares to financial institutions and private investors. Conflicts can ensue when the dominant founder of a company wants to push it into new areas that investors are uncomfortable with. Problems also arise when the company board, often controlled by major investors, decides that the founder is no longer suited to run the firm. Steve Jobs was forced to quit Apple after a boardroom battle in the 1980s. The same fate befell John McAfee (1946–), the inventor of antivirus software. After taking his company public and making a huge personal fortune, McAfee left, allegedly after a boardroom disagreement.

Some inventors choose to avoid being pushed out of their companies by moving to more advisory roles soon after their companies are formed. After setting up eBay, Pierre Omidyar quickly hired an experienced business manager to help him turn it into a leading corporation. The founders of Yahoo! and Google also brought in experienced managers very early on, though they remain actively involved in guiding the businesses they established.

Inventors also face difficulties when they do not have the resources to develop their ideas and are forced to sell them to a company in return for a cut of the profits. Chester Carlson (1906–1968), inventor of the photocopier, was unable to sell his idea to IBM, RCA, or General Electric, none of which could see the potential in office copiers. Eventually, Carlson found a small photographic company, Haloid, willing to manufacture and promote the photocopier. Haloid (which later changed its name to Xerox) made a great success of the invention, and Carlson earned royalties from the copiers Haloid developed. Xerox continued its reputation for corporate invention at its Palo Alto Research Center (PARC) in the early 1970s (see box, PARC).

WHY CORPORATIONS DOMINATE

Corporations appear to be better suited than individuals to originate and develop inventions. The most obvious reason is the greater resources of large corporations; they have more money to spend on research and a wider pool of talented people to work on problems. Established corporations also typically have their own patent and legal departments to protect their ideas. In some areas, research has become so complex and expensive that individuals have no hope of competing with corporations. A lone inventor

PARC

On July 1, 1970, Xerox Corporation officially opened
its new laboratory in Palo Alto, California. Called Xerox
PARC (Palo Alto Research Center), the laboratory developed most of the crucial technologies used in modern personal
computers in a burst of creativity during the early 1970s.

Xerox had made its fortune in the 1960s by developing the photocopier, but office computers had also started to become extremely popular
during the same period. This led Xerox executives to believe that offices
might eventually become "paperless," making their photocopiers obsolete
and threatening their core business. Consequently, they thought it wise to
explore making computers themselves—and PARC was born.

One of PARC's first successes was the invention of the laser printer in
1971. The following year, PARC introduced object-oriented programming, a way of building up a computer's instructions out of self-contained
modules. These modules can then be reused in other programs, saving
time and effort. A great invention in 1973 was the Xerox Alto, a user-
friendly computer with a picture-based, desktop screen.

Many corporations benefited from PARC's research in the 1980s.
Hewlett-Packard, for example, made a fortune developing laser printers:
Apple and Microsoft developed graphical user interfaces. Xerox had
borne the cost of developing these inventions, but other companies prof-
ited from them far more. Ever since, critics have questioned why Xerox
had done so much to invent personal computing in the early 1970s—and
so little to profit from its inventions thereafter. Some believe PARC's
inventions were simply too far ahead of their time. Others have argued
that PARC produced so many
ideas so quickly that Xerox did
not have the resources to
develop them properly. Others
think Xerox, as a photocopier
manufacturer, could not appreciate the value of inventions it
had created outside its main
area of business.

The offices of Xerox PARC.

could not compete with Intel to develop a faster microchip or with Ford to develop a more efficient car engine; both projects would take many millions of dollars and years of research.

Over time, companies develop many strengths. They build up large teams of experienced scientists and engineers, so their fortunes are seldom dependent on even the most talented individuals. Company names also become familiar brands, which can help them sell new ideas. During its heyday in the 1960s and 1970s, IBM, the world's biggest computer company at the time, was immensely profitable partly because of a well-known saying that "No one ever got fired for buying IBM." Corporate research and development labs can be attractive places for people to work because of their reputations and higher salaries.

As companies grow, however, they can lose their flexibility and willingness to innovate. The Ford Motor Company started experiencing losses in the 1930s for this reason, as did IBM in the 1980s and 1990s. Some large corporations try to keep themselves innovative by using high-tech research and development laboratories. In such places, creative engineers and scientists are encouraged to develop promising new "blue-sky" ideas, sometimes without any pressure to produce successful products.

Corporations have dominated invention since the early 20th century. As science becomes ever more specialized, large universities now often lack the facilities for turning new discoveries into practical technologies, and wealthy companies with big research departments are generally more innovative. Yet important ideas can still be developed by large communities of people without corporate help. The World Wide Web is an example of a great invention that was developed in this way. Indeed, its inventor, Tim Berners-Lee (1955–), has always been determined that the Web will not be taken over by corporations and has refused to patent or profit directly from his idea. While most patents are now filed by corporations, success stories like the Web, and the recent triumph of companies like eBay, Yahoo!, and Google, confirm that a place still exists for individuals in the future of invention.

—Chris Woodford

Further Reading

Bender, Lionel. *Eyewitness: Invention.* New York: Dorling Kindersley, 2005.

Gehani, Narain. *Bell Labs: Life in the Crown Jewels.* Summit, NJ: Silicon Press, 2003.

Hiltzik, Michael. *Dealers of Lightning: Xerox PARC and the Dawn of the Computer Age.* New York: Collins, 2000.

See also: Bardeen, John, Walter Brattain, and William Shockley; Bell, Alexander Graham; Benz, Karl; Berners-Lee, Tim; Carlson, Chester; Carothers, Wallace; Caselli, Giovanni; Cooper, Martin; Deere, John; Eastman, George; Edison, Thomas; Ford, Henry; Goodyear, Charles; Jobs, Steve, and Steve Wozniak; Kapany, Narinder; Kilby, Jack, and Robert Noyce; Kwolek, Stephanie; McAfee, John; McCormick, Cyrus Hall; Omidyar, Pierre; Plunkett, Roy J.; Schawlow, Arthur, and Charles Townes.

JACQUES-YVES COUSTEAU

Inventor of the Aqua-Lung
1910–1997

Oceanographer, writer, filmmaker, and environmentalist, Jacques Cousteau is known to a generation of readers and television viewers around the world as a passionate explorer of the world's oceans and rivers. As captain of the *Calypso*, Cousteau brought audiences with him on his journeys, creating some of the most breathtaking underwater footage ever recorded. He was the first person to share the beauties of the undersea world with a global population, to argue for the preciousness of the oceans, and to warn of the dangers human irresponsibility posed to them. He was able to become a pioneer of undersea observation because of one invention: in 1943, with French engineer Émile Gagnan, Cousteau developed the Aqua-Lung, also known as the self-contained underwater breathing apparatus, or "scuba."

EARLY YEARS

Jacques-Yves Cousteau was born June 11, 1910, in Saint-André-de-Cubzac, a small town in the southwest of France. The son of Daniel Cousteau and Elizabeth Duranthon, Cousteau traveled extensively with his family while his father worked as business manager and legal adviser for a wealthy American. In 1920 the family moved for a time to New York City, where "Jack" Cousteau attended Holy Name School. During his two years in the United States, he learned to speak and write fluent English.

In 1930, back in France, Cousteau entered the naval academy, where he became a gunnery officer. While training to be a pilot, he was involved in a car accident that left him badly injured and ended his hopes of a career in aviation. He turned his attention instead to the sea. As Cousteau wrote in his book *The Silent World* (1953), he experienced a kind of revelation when, in 1936, he went swimming underwater near the port of Toulon. "I was astounded by what I saw . . . rocks covered with green, brown and silver forests of algae and fishes unknown to me, swimming in crystalline water." In 1937, Cousteau married Simone Melchoir, who also became an accomplished diver. They had two sons who would also become avid divers and marine enthusiasts.

Cousteau was eager to enter more fully into the world he had discovered in 1936. He began skin diving, but the amount of time he could spend underwater was limited to the amount of the time he could hold his breath. Cousteau rejected helmet diving, which allowed a diver to remain underwater for longer periods by connecting through a hose to an oxygen source on a ship. Immobile, with extremely limited vision, a helmet diver is, in Cousteau's words, "a cripple in an alien land."

> Sometimes we are lucky enough to know that our lives have been changed. It happened to me that summer's day [in 1936] when my eyes opened to the world beneath the surface of the sea.
>
> —Jacques Cousteau

Jacques-Yves Cousteau, pictured in 1956.

THE AQUA-LUNG

Cousteau was intent on overcoming the obstacles of traditional diving equipment. With his friend, engineer Émile Gagnan, he began exploring methods for self-contained underwater breathing. The two developed an automatic regulator capable of providing compressed air on demand. Combined with a hose that connected a mouthpiece to two tanks of compressed air, this regulator proved effective. Cousteau tried out the prototype Aqua-Lung in 1943; later that year the two men were granted a joint patent.

Cousteau was elated at the freedom of movement the new gear allowed him. "I experimented with all possible maneuvers—loops, somersaults, barrel rolls. . . . Delivered from gravity and buoyancy, I flew about in space," he wrote. He continued to refine the mechanism as he undertook several underwater expeditions to find ships that had sunk during World War II. He also started photographing underwater life.

In 1950, he acquired the *Calypso*, which had been used as an American minesweeper during the war, and converted it into an oceanographic vessel. Equipped with an array of nautical and scientific instruments, *Calypso*

Émile Gagnan

The charismatic adventurer Jacques Cousteau is known to millions, but the man with whom he developed the Aqua-Lung remains virtually unknown. Émile Gagnan was born in France in 1900 and graduated from technical school in the early 1920s. In 1943, Gagnan was working on a regulator for automobiles when Cousteau enlisted his help in designing underwater breathing equipment. The result of their collaboration was a demand-valve that could deliver an air supply as needed from a tank to the diver's mouthpiece. This valve, originally used to regulate the gas supply in combustion engines, was the perfect piece for a self-contained underwater breathing apparatus, or "scuba."

The first scuba equipment, referred to as the "Aqua-Lung," appeared commercially in France in 1946. The following year, Gagnan moved with his family to Montreal, Quebec, where he took a job with Canadian Liquid Air Limited. He continued designing and developing scuba technologies, making important contributions to the advanced scuba equipment in use today.

sailed into nearly all the waters of the world, amassing along the way a stunning visual record and carrying out much valuable scientific research.

COUSTEAU THE FILMMAKER

Cousteau and his crew gained recognition beginning in the mid-1950s with their documentaries *The Silent World* (1956), *World without Sun* (1964), and others. In 1966, Cousteau and *Calypso* began to be household names when an hour-long documentary special, *The World of Jacques Cousteau*, aired on American television. Narrated by actor-director Orson Welles, the program was extremely popular.

Two years later, *The Undersea World of Jacques Cousteau* began its eight-year run on television. With his sons Philippe and Jean-Michel, Cousteau filmed marine life around the globe and brought the stunning results into millions of homes. Narrated in Cousteau's distinctly poetic style, the programs' evocative titles, "The Savage World of the Coral Jungle" (1968), "The Smile of the Walrus" (1972), and "The Incredible March of the Spiny Lobsters" (1976), for example, give some idea of the lyrical qualities of Cousteau's underwater vision.

The Undersea World of Jacques Cousteau broadcast its final program in the spring of 1976. A new series, *Cousteau Odyssey*, began airing the following year. Focusing more on Cousteau's growing environmental concerns than on particular marine species, this new series produced a dozen installments that ran through 1981. Other documentary work for television included the four-part *Cousteau/Amazon* (1984–1985), which highlighted the threatened environment and cultures of the Amazon region.

Scuba Gear

scuba mask

mouthpiece

Cousteau's invention, the regulator

BCD (Buoyancy Compensation Device)

scuba tank

wet suit

swim fins

The basic components of scuba diving equipment. Cousteau's invention, the regulator, has two parts: the first attaches to the air tank and reduces the pressure of the air flowing from the tank. This depressurized air then flows through a hose to the second part, connected to the diver's mouthpiece, which reduces the air pressure even more. The diver can then breathe normally.

Cousteau and his colleagues adjust his equipment while preparing for a dive.

OTHER INVENTIONS

While sailing the world's seas, Cousteau continued to develop new technologies to better observe and film the multitude of life-forms he found. In the early 1950s, he assisted Swiss physicist Auguste Piccard in the development of the bathyscaphe, a deep-diving underwater vehicle. Cousteau also introduced new technologies in underwater filming equipment, vastly improving the quality of the images he was able to bring to his millions of viewers.

In 1959, he invented with Jean Mollard a mini-submarine—an easily maneuvered two-person diving machine capable of descending to considerable depths. Around that same time, Cousteau experimented

TIME LINE

1910	1943	1950	Early-1950s	1959
Jacques-Yves Cousteau born in Saint-André-de-Cubzac, France.	Cousteau and Émile Gagnan develop and patent the Aqua-Lung.	Cousteau acquires the *Calypso*.	Cousteau assists Auguste Piccard in development of the bathyscaphe.	Cousteau invents a mini-submarine with Jean Mollard.

1966	1968–1976	1977–1981	1997
The World of Jacques Cousteau airs on American television.	*The Undersea World of Jacques Cousteau* airs on television.	*Cousteau Odyssey* airs on television.	Cousteau dies.

with a device called the "sea flea," a single-diver pod-shaped vehicle that could descend to depths of about five hundred feet. His greatest achievement as an inventor, however, remained his first: the Aqua-Lung revolutionized undersea study and gave birth to the widely popular hobby of scuba diving.

COUSTEAU'S IMPACT

The Aqua-Lung allowed Cousteau to record the images and impressions that hypnotized his audience, but Cousteau's unique personality, vision, and passion for marine life are what make his work special and his legacy

Cousteau explains the "sea flea" to French reporters in 1967.

lasting. That passion led Cousteau increasingly to speak out against the destruction of ecosystems—both at sea and on land. In 1973 he established the Cousteau Society, a nonprofit organization devoted to preserving the environment through better management. The society continues to operate toward this goal into the 21st century.

Cousteau garnered numerous awards and citations over his long career. His 1956 film *The Silent World* won the Palme d'Or at the Cannes International Film Festival as well as an Oscar from the American Academy of Motion Picture Arts and Sciences—just two of the many awards he received for his dozens of books and movies

Watchdog of the Planet

To millions of viewers and readers, the words *Calypso* and *Cousteau* are intimately linked. Cousteau captained his beloved ship into nearly every corner of the world's waters and introduced large numbers of people to the beauties he discovered. *Calypso*, however, was not the only ship that occupied the captain's thoughts. Cousteau's commitment to improving the environment led him to consider alternatives.

In the early 1980s Cousteau conceived of a new kind of high-tech sail to be used on large ocean ships to help reduce fuel consumption. Called Turbosails, two of these 35-foot (10.7-m) high sails were used to outfit the *Alcyone*, which was launched in 1985 and remains the expedition ship of the Cousteau Society. According to the society, the *Alcyone* "celebrates . . . the wedding of hydrodynamics and aerodynamics" and cuts fossil-fuel use by up to 35 percent.

In 1994, the *Calypso* sank after being accidentally hit by a barge in Singapore Harbor. Cousteau set about designing *Calypso II*. Intended to incorporate television studio equipment with marine laboratory equipment, *Calypso II* was dubbed, according to the Cousteau Society, "Watchdog of the Planet."

Cousteau's ship, the Calypso, *moored in Marseilles, France.*

spanning some fifty years. In 1959 he addressed the inaugural World Oceanic Congress; this occasion brought him wide recognition and contributed to his appearance on the cover of *Time* magazine on March 28, 1960. The following year, Cousteau was awarded the Gold Medal of the National Geographic Society by President John F. Kennedy. He was later awarded the French government's Legion of Honor and the U.S. Presidential Medal of Freedom, both in 1985. Cousteau died in Paris on June 25, 1997, at the age of 87.

—Paul Schellinger

Further Reading

Book
Munson, Richard. *Cousteau, the Captain and His World.* New York: Morrow, 1989.

Web site
Cousteau Society Web site
> Biography of Cousteau and information about the society's mission.
> http://www.cousteau.org

See also: Science, Technology, and Mathematics.

BARTOLOMEO CRISTOFORI

Inventor of the piano
1655–1731

While working as a harpsichord maker in Florence, Bartolomeo Cristofori invented the piano. This new instrument differed from the harpsichord in that it featured a keyboard connected to a number of hammers that would strike a row of strings to sound various notes. The harpsichord, by contrast, produced sound by a series of picks operated by the keyboard that would pluck a bank of strings. Cristofori's new technology allowed for greater flexibility in tonal quality and introduced variation in dynamics (volume of sound) and set the piano on its course to becoming a major instrument on concert stages in the 19th century.

EARLY YEARS

Bartolomeo di Francesco Cristofori was born in Padua, in northern Italy, on May 4, 1655. He was active as an instrument maker in Padua until 1688, when he was brought to the court of Prince Ferdinando de Medici in Florence. There he was employed as a maker of various keyboard instruments, including the harpsichord and the spinet. The prince, who already had an impressive collection of harpsichords, hired Cristofori to maintain his collection and build new instruments.

The harpsichord had been a much favored keyboard instrument since its invention in Flanders in the early fifteenth century. It consisted of strings stretched over a wooden sounding board; a row of picks connected by wires to the keyboard would pluck these strings, producing different notes. Although it proved capable of producing complex sounds and beautiful music, the harpsichord was unable to make gradations in the dynamics of sound. Whether the player struck the key with great force or very little force, the sound stayed at a constant volume. Over time, some harpsichords were developed with three or more keyboards connected to separate banks of strings; these instruments were capable of producing greater depths of sound, but they were still relatively limited.

INVENTS THE PIANO

Even after the invention of the piano, the harpsichord remained a popular instrument in the 18th century. This 1764 portrait shows a young Wolfgang Mozart playing a harpsichord; his father, Leopold Mozart, plays violin, while sister Maria Ann sings. (Lithograph by A. Schieferdecker, based on a watercolor by Louis Carrogis de Carmontelle.)

The year given for Cristofori's invention of a new kind of keyboard instrument that struck, rather than plucked, wires to produce sound is 1709. However, records from the period indicate that in 1700, an inventory of Medici instruments in Florence included an "*arpicebalo* newly invented by Bartolomeo Cristofori." This instrument resembled a harpsichord (the Italian is literally translated as "harp-harpsichord") and featured hammers capable of striking the strings from below. These hammers, either of wood or of paper covered by deerskin, could be made to strike a particular string with greater or lesser

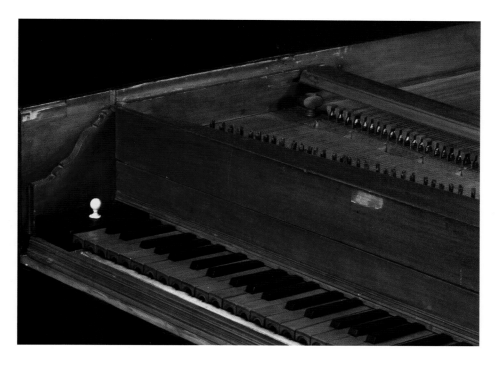

A grand piano designed by Cristofori.

force, producing louder or softer tones, depending on the pressure exerted on that key.

Later, in a journal article published in 1711, Scipione Maffei would describe Cristofori's instrument as a *gravicembalo col piano e forte*—a "clavichord [similar to a harpsichord] that plays soft [piano] and loud [forte]." This was the first reference to the instrument's ability to produce different volumes of sound. The name was shortened to "pianoforte" and remained in use through the first few decades of the 19th century. Eventually the instrument became known simply as the piano.

Cristofori spent several years improving his invention. Contemporary sources indicate that four pianofortes existed by 1711. After his patron Ferdinando de Medici died in 1713, Cristofori remained in Florence, where he continued to improve the technology of his new piano. By 1726 Cristofori's piano had all the elements of the modern piano, including a check that would hold the hammer in place after it struck a wire. This mechanism would prevent the hammer from bouncing back and striking the wire a second time. Cristofori made an instrument in which each hammer struck two wires at the same time. He then installed an early version of what would become the modern soft pedal: by manually moving one of two knobs on either side of the keyboard, the player could make the hammer strike just one of the wires. (The soft pedal on modern pianos is underneath the keyboard and is operated by foot.) This was called playing *una corda* (one string) and remains a common marking in piano music scores.

A QUIET RECEPTION

This 1910 illustration by Edward Gooch shows the evolution of the piano from the spinet (top left), to the harpsichord (top right), to various styles of pianos. Seated at each of the instruments is a composer who was particularly associated with that instrument.

Yet for all his advancements, Cristofori's pianoforte did not gain much favor. Since Cristofori had fashioned his instruments entirely of wood, he could not stretch the wires very tightly without compromising the integrity of the structure of his instrument. Consequently, he could not create enough tension to allow the instrument to play very loudly. Many of Cristofori's contemporaries considered the pianoforte too weak in its sound production and continued to prefer the harpsichord. (The modern piano has a metal frame that permits wires to be strung very tightly, thus achieving much greater tension and ability to create significant volume.)

Cristofori's invention received little notice in Italy during his lifetime and in the years following his death in Florence in 1731. In Germany, however, it gained wider recognition. This may have been a result of a translation of Maffei's article into German in 1725; the translation was included in a collection entitled *Critica musica* by Johann Mattheson. Maffei's article, which featured a diagram of Cristofori's design and depicted the action of his pianoforte, was apparently read by a leading organ builder in the Saxon court, Gottfried Silbermann. Beginning in the 1730s, Silbermann began to make his own versions of Cristofori's piano. Even into the late 18th century, the invention of the

piano was being attributed not to Cristofori but to Silbermann, who appears to have encouraged this confusion.

The composer Johann Sebastian Bach (1685–1750) played one of Silbermann's instruments and originally dismissed it before eventually praising a later version. Bach helped to spread Silbermann's reputation

The Piano and the Romantic Movement

In the late 18th century, an artistic movement in Europe emerged known as Sturm und Drang—literally, storm and stress. This movement, which emphasized the emotional involvement of the artist and the viewer, listener, or reader in creating and experiencing a work of art, was an important precursor to European romanticism. Romanticism explored the emotional and subjective possibilities of art in a truly unparalleled way in the early to mid-19th century.

The piano suited this temperamental shift extremely well. It had been developed at a time when the expressive range of musical forms was expanding: opera, oratorio, the symphony, chamber music, and concertos began to flourish in the seventeenth century and the early 18th century. A new, more flexible, more expressive kind of keyboard instrument was needed to suit the flowering of musical styles.

By the early 19th century, music for solo piano was being composed by Wolfgang Amadeus Mozart, Franz Josef Haydn, and Ludwig van Beethoven, to name just three of the greatest early composers for the piano. Compositions for the piano became as well known as operas or symphonies. The emergence at the same time of a stronger middle class throughout Europe and America meant that more and more people could afford to purchase pianos for their homes, ensuring the piano's central place in the world of musical performance, entertainment, and education.

Leading Romantic composer Ludwig van Beethoven is depicted at work in this portrait from around 1800 (creator unknown).

TIME LINE

1655	1688	1709	1711	1726	1731
Bartolomeo di Francesco Cristofori born in Padua, Italy.	Cristofori employed by Prince Ferdinando de Medici's court in Florence.	The year given for invention of Cristofori's new keyboard instrument.	The first reference to the instrument appears in a journal article.	Cristofori builds an instrument similar to the modern piano.	Cristofori dies in Florence.

as the "inventor" of the piano. Nevertheless, modern scholars have determined that Silbermann drew his inspiration almost exclusively from Cristofori's designs.

THE MODERN PIANO

The qualities of Cristofori's instrument became more apparent in the late 18th century and especially in the 19th century, when the piano began to assume its role as chief solo instrument (played by itself, with other solo instruments, or accompanied by an entire orchestra) on concert stages around the world and in people's living rooms. Cristofori's keyboard instrument—capable of playing soft or loud and offering a wide range of tonal qualities—appeared in average people's homes, unlike harpsichords, which had mostly belonged to royalty and to the wealthy.

The modern piano has 88 keys, in contrast to Cristofori's, which had between 54 and 60 keys. By the beginning of the 20th century, the piano was embraced throughout the world as the primary tool for musical education and, some might say, the most popular of all instruments.

—Paul Schellinger

Further Reading

Book
Pollens, Stewart. *The Early Pianoforte.* Cambridge: Cambridge University Press, 1995.

Web site
Piano Education Page
> Wide range of information about the piano, including its invention and history. http://pianoeducation.org/index.html

See also: Entertainment; Paul, Les; Sax, Adolphe.

LOUIS DAGUERRE

Inventor of the photograph
1789–1851

Ever since cave dwellers painted pictures of the animals they hunted, people have been trying to capture images of things that were important to them. Until the mid-19th century, the only way to record a scene was to paint or draw it. When French artist Louis Daguerre announced the invention of photography in 1839, he changed history—by making history itself easier to capture in pictures.

EARLY YEARS

Louis-Jacques-Mandé Daguerre was born in Cormeilles-en-Parisis near Paris, France, in 1789. Little is known about his early life and education. Although he started working as a tax collector, he was soon earning his living in more creative ways. He worked as an architect's apprentice for a time and then, at age 16, became a scene painter and set designer. From 1819 to 1822, he was set designer at the Paris Opera, where he painted huge landscape backdrops and designed unique lighting effects. *Aladdin or the Marvelous Lamp* was a notable production in 1822, in which Daguerre was the first to use gas lamps to light the stage.

Daguerre's theater work inspired his first business idea, the Diorama, in 1822. Working with his friend Charles-Marie Bouton, he painted enormous pictures of dramatic or famous events on translucent paper, typically 45 feet by 71 feet (13.7 m by 2.16 m). Then he positioned lights around and behind the pictures, switching the lights on and off to suggest changes in the weather or time of day. Adding music, animals, actors, props, and special effects, he created breathtaking scenes that people paid to view. The scenes Daguerre painted had titles like *Land-Slip in the Valley of Goldau* and *Cathedral of Sainte Marie de Montréal* and were based on real-life places and happenings.

In order to be effective, the Diorama scenes had to be painted as realistically as possible. This led Daguerre to wonder if there were better ways of copying the things he saw.

Louis Daguerre, photographed using his own process, around 1845.

PARTNERS IN PHOTOGRAPHY

The very same question had already occurred to Joseph-Nicéphore Niépce (1765–1833). Beginning around 1814, Niépce had tried to find ways of capturing sunlight on paper. He began by using a simple device called a camera obscura that artists had been using since at least 1500. This device can be a large darkened room with a

An audience views Daguerre's Diorama (undated illustration).

tiny hole in the drapes or a huge sealed box or chamber with a small hole in one wall. When light enters through the hole, it throws an image of the outside world onto the opposite wall inside. There was nothing in a camera obscura that could record light; instead, artists would hang pieces of paper on the wall and then sketch around the image they saw to record it. However, during the 18th century, several chemists, including the German Johann Schulze (1687–1744) and the Swede Carl Scheele (1742–1786), discovered that some chemicals based on silver changed color when they were left out in the light. This discovery opened up the possibility of using chemicals in a camera obscura to record images.

In 1827, Niépce combined these two ideas and made the world's first photograph. First, he prepared what was called a photographic plate—a piece of metal (pewter) covered in a light-sensitive chemical (bitumen, a kind of asphalt). Then he put the plate upright inside a camera obscura and left it by the window of his house. Very slowly, over about eight hours, the plate captured a fairly blurred image of the view from his window (a process known as exposing the plate to light). Niépce called his invention heliography (which means "drawing by the sun").

Daguerre had also started experimenting with chemicals to see if he could capture sunlight permanently on paper. He learned of Niépce's work and wrote to him on December 14, 1829. The two men became partners, hoping they would develop a quick and easy way of capturing

sunlight on paper. When they switched to using copper plates and a light-sensitive chemical based on iodine, they managed to cut the exposure time from eight hours to 20 or 30 minutes, but otherwise their results were not much better.

BIRTH OF THE DAGUERREOTYPE

After Niépce died in July 1833, Daguerre continued his experiments with the help of Niépce's son, Isidore. A couple of years later, he made a dramatic breakthrough and discovered a better way of capturing light on photographic plates. There is a story that Daguerre threw some partly exposed plates into a closet where there was a broken thermometer. When he looked at the plates a few days later, he found an image had formed on them. Daguerre realized that the mercury from the thermometer was the cause: the vapor it had given off had somehow "fixed" the image onto the metal plates. However, historians of photography think this story was probably made up later; no one really knows how Daguerre discovered the importance of mercury.

However it happened, Daguerre had arrived at a much better way of taking pictures. It involved several stages. First, he took a copper plate, covered it in silver, and treated it with iodine vapor. This caused a chemical reaction that coated the plate with silver iodide, a light-sensitive compound. Daguerre put the plate in his camera and exposed it to light.

Next, he "developed" the plate (revealed the image it contained) by letting fumes from hot mercury pass over it. Finally, he "fixed" the developed image (made it permanent) by treating it with common salt. The whole process was much quicker than anything that had been previously used.

Isidore Niépce was delighted. On November 1, 1837, he wrote to Daguerre: "What a difference . . . between the method which you employ and the one by which I toil on! While I require almost a whole day to make one design, you ask only four minutes! What an enormous advantage! It is so great, indeed, that no person, knowing both methods, would employ the old one." Niépce realized that Daguerre had made a breakthrough and, as a result, allowed his partner to name their new way of recording pictures the daguerreotype process.

PAINTING IS DEAD

If Louis Daguerre and his new partner Isidore Niépce thought they were going to make their fortune, they were soon disappointed: when Daguerre tried to sell their invention, no one wanted to buy it. Luckily, there was interest of a different kind. On January 9, 1839, Daguerre demonstrated his invention at a meeting of the French Academy of

The invention of photography had a revolutionary impact on the way history—be it political or personal—was recorded. The Benjamin family was photographed around 1850.

President Abraham Lincoln, photographed in 1864.

Unidentified young girl, photographed in 1847.

A tinted daguerreotype advertising the candidacy of Franklin Pierce, who was president of the United States from 1853 to 1857.

Unidentified woman, photographed in the 1850s.

Sciences; the audience was astonished by the incredible detail in the pictures he showed. Around this time, Daguerre's invention caught the attention of François Arago, a respected scientist and politician. He saw the potential immediately and came up with a novel suggestion: the French government should buy the rights to the daguerreotype and "then nobly give to the whole world this discovery which could contribute so much to the progress of art and science."

On June 15, 1839, the government set up a commission to consider the idea. Chaired by painter and art expert Paul Delaroche, the commission unanimously endorsed the plan about six weeks later. One said the daguerreotype was a "beautiful discovery." Delaroche saw it as the beginning of a whole new era in the visual arts, boldly declaring: "From today, painting is dead."

The equipment necessary to create a daguerreotype is pictured here. The silver-coated plate (center) was polished with a buffer (right). After polishing, the plate was placed in a sensitizing box (top) where it was exposed to iodine and bromine vapors. After the plate was exposed and an image was captured on the plate, the daguerreotype was developed with heated mercury in the developing box (left). Finally, the beveller (bottom) would be used to put an edge on the plate.

Talbot and the Negative

Photography turns everyone into an artist: few people can paint a good portrait of a friend or a landscape that they love, but anyone can take a photograph. Louis Daguerre was inspired to invent the daguerreotype when he realized he could not draw quite as well as he would have liked. Exactly the same reasoning inspired the Englishman who invented the modern photographic process. William Henry Fox Talbot was sketching lakes in Italy when he decided there had to be a better way of recording nature.

Unlike Louis Daguerre, Talbot tried to record images on paper, not metal, but he used silver compounds to capture light in a very similar way. He made his photographic paper by soaking it in common salt (sodium chloride) and then silver nitrate. A chemical reaction happened on the paper, coating it with light-sensitive silver chloride. Talbot placed the now-sensitive paper in the back of a camera, opened the lens to light for a minute or so to take a picture, then covered the lens again. Unlike the daguerreotype process, Talbot's process burned a ghostly negative image onto the paper: dark areas in the original picture showed up light and vice versa. Talbot then took the negative into a darkened room and treated it with other chemicals to produce a positive print in which the original light and dark areas were now reversed and showed up correctly.

Talbot announced his invention in 1839, the same year the French government made public Daguerre's process. The two men immediately became rivals. The best thing about Talbot's photographs (called Talbotypes or calotypes) was that the negative could be used to make any number of positive prints—something that was not possible with daguerreotypes. Whereas daguerreotypes were pin-sharp, Talbot's method produced softer and more artistic prints that many people thought too blurred. Daguerreotypes were free for anyone to use, thanks to the French government, but Talbot patented his process and tried to charge people for using it.

Daguerre gained the early advantage, but Talbot's process—and the improved wet-plate process developed by Frederick Scott Archer—eventually triumphed and evolved into the photographic film process still used today. It is ironic that Louis Daguerre remains more famous than William Henry Fox Talbot. Daguerre's original, metal-plate photographs have lasted longer and are now considered to be important historical records.

According to the terms of the deal, the French Academy of Sciences gained the rights to the daguerreotype and made the invention free for anyone to use. Accordingly, Daguerre wrote a detailed description of the invention—*An Historical and Descriptive Account of the Various Processes of the Daguerreotype and the Diorama*—which was published that August. In return for giving his invention to the government, he was made an officer of the Legion of Honor (a prestigious French award). He was also given a modest annual payment of 6,000 francs (equivalent to about $21,000 today), and Isidore Niépce was given 4,000 francs (equivalent to $14,000 today) to recognize his father's important work.

Once Daguerre's invention was made available to all, other inventors looked at ways of improving the process. Daguerre himself did little more. Instead, he returned to his first love, painting, and enjoyed a long retirement on his government pension. When he died at Brusur-Marne, France, on July 10, 1851, an obituary in the *Illustrated London News*, an English weekly, noted: "He was a man of extreme modesty and great personal worth, and devoted to his profession, that of an artist."

THE RISE OF PHOTOGRAPHY

The impact of Daguerre's invention was immediate. His booklet was an instant, international best seller: by the end of 1839, it had been published in a dozen languages in 29 different editions. In the United States, Daguerre's process was popularized by the artist Samuel Morse (1791–1872), who became even more famous as the inventor of the electric telegraph. The process was later improved by scientist John William Draper (1811–1882). It spread rapidly when many people began to set themselves up as photographers specializing in daguerreotype, offering to capture portraits on shiny metal plates for around two

TIME LINE

1789	1814	1819	1822	1827
Louis Daguerre born in Cormeilles-en-Parisis, France.	Joseph-Nicéphore Niépce begins his work on photography.	Daguerre becomes a set designer at the Paris Opera.	With Charles-Marie Boulton, Daguerre creates the Diorama.	Niépce creates the first photograph.

A teenager shows her family photographs taken with a cell phone.

to five dollars each. Millions of these cameo portraits had been taken by the 1850s.

Although the daguerreotype remained popular for about 20 years, it was eventually overtaken by better methods of photography. In England, William Henry Fox Talbot (1800–1877) developed the mod-

TIME LINE

1829	1833	1833	1839	1851
Daguerre contacts Niépce, and they become partners.	Niépce dies; Daguerre continues his experiments.	Daguerre creates the first daguerreotype.	Daguerre demonstrates his daguerreotype process to the French Academy of Sciences to great acclaim.	Daguerre dies.

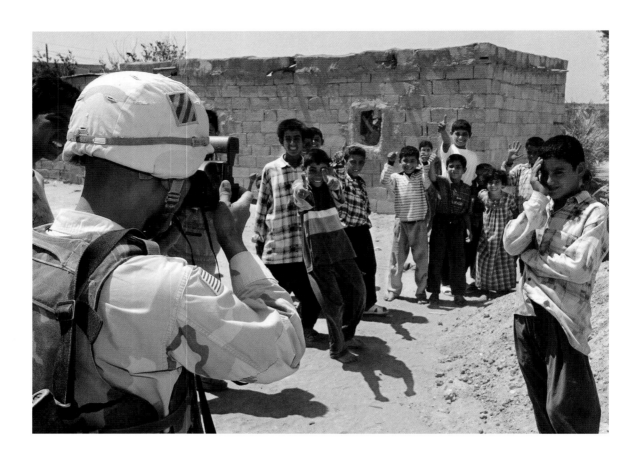

ern photographic negative process (see box, Talbot and the Negative). This allowed many copies to be made from an original image, something that was not possible with the daguerreotype. Talbot's method took a long time to catch on, and it was improved, in turn, by another English scientist, Frederick Scott Archer (1813–1857). Instead of using metal plates covered with dry chemicals, Archer coated his wet plates with liquid chemicals; although messier, they produced sharper images much more quickly and led eventually to the modern photographic film, developed by George Eastman (1854–1932) in 1883.

Together, these pioneers of photography made history by changing the way history was recorded: until the invention of photography, history was mostly about the written word and the painted picture. Photography was something entirely different: it actually captured moments of history, as people saw them and as they were happening. Thanks to Niépce, Daguerre, and their colleagues, humankind now has a photographic record of many important events, including photographs from when astronauts first set foot on the moon and when the Berlin Wall was knocked down. In addition to recording significant events in human history, photographs also allow people to capture memories of their own lives and families. Today, photography is everywhere—in books, newspapers, magazines, and billboards—and is taken entirely for granted. Photography also surrounds us in other ways: at the end of the 19th century, the Frenchmen Auguste and Louis Lumière invented a camera that turned still photographs into movies, which led to the invention of television.

Louis Daguerre did not invent photography by himself. His French colleague Niépce had already developed half the process, and his English rival Talbot independently arrived at a very different way to create photographs. Yet much of the early excitement about photography was caused by the daguerreotype, for several reasons. The sharpness of its images and the way they captured a "likeness" sold people on Daguerre's idea; only the rich could have portraits painted, but most people could own a daguerreotype. Although many different stages were involved in taking a daguerreotype, they were fairly easy to learn; in fact, one of the original claims made about Daguerre's invention was that "anyone may succeed . . . and perform as well as the author of the invention."

Another key to Daguerre's success was the French government's decision to buy his invention. Some inventors might have refused, believing they could make more money by selling their idea themselves. By settling for a more modest reward, Daguerre allowed the French government to give the process of photography as a gift to the world—a gift that has preserved memories and enriched lives ever since.

—Chris Woodford

Paparazzi at the 2006 Cannes Film Festival in France. Photography has been a major contributor to the development of celebrity culture in the 21st century.

Further Reading

Books

Bankston, John. *Louis Daguerre and the Story of the Daguerreotype*. Hockessin, DE: Mitchell Lane, 2004.

Johnson, Neil. *Photography Guide for Kids*. Washington, D.C.: National Geographic, 2001.

Rosenblum, Naomi. *A World History of Photography*. New York: Abbeville, 1997.

Web sites

American Museum of Photography
Extensive information about the history of photography.
http://www.photography-museum.com/

Daguerre Society
A society whose members study the history of the daguerreotype.
http://www.daguerre.org/

Library of Congress Daguerreotype Collection
American history recorded in early American daguerreotype photographs.
http://memory.loc.gov/ammem/daghtml/daghome.html

The Making of a Daguerreotype
A detailed look at the process involved.
http://www.ccsd.ca/charlotte/dagazine/mi/exhibit/brochure.htm

See also: Communications; Eastman, George; Edison, Thomas; Entertainment; Lumière, Auguste, and Louis Lumière; Morse, Samuel.

GOTTLIEB DAIMLER

Pioneer automobile manufacturer, inventor of the motorcycle

1834–1900

The name Gottlieb Daimler, little known to Americans since his death, became more widely recognized when the company he founded, the Daimler Motor Company (which would produce the Mercedes-Benz automobile), merged with the American Chrysler Corporation in 1998 to form DaimlerChrysler. Daimler's improvements to the internal combustion engine laid the groundwork for the automobile industry that would arise in the early years of the 20th century.

EARLY YEARS

Gottlieb Wilhelm Daimler was born March 17, 1834, in Schorndorf, Württemberg (near Stuttgart), Germany. The son of a master baker, Daimler attended technical school from 1848 to 1852, while also apprenticing with a gunsmith in Stuttgart. From 1853 to 1857, he worked in a steam-engine locomotive factory in Strasbourg. He then completed his education at the Stuttgart Polytechnic, in preparation for a career as a mechanical engineer.

After working briefly in Paris, Daimler traveled in England in 1861, working at a variety of engineering plants in Leeds, Manchester, and elsewhere. In 1862 he attended the world's fair in London, where he saw the latest advances in European engineering.

DAIMLER AND MAYBACH

Daimler returned to Germany in 1863 and became manager of an engineering workshop in Reutlingen. This facility, known as the Bruderhaus, operated in conjunction with a progressive orphanage designed to provide orphans and other socially disadvantaged children with the opportunity to learn engineering skills. Among the orphans living at the Bruderhaus, Daimler met the teenage Wilhelm Maybach (1846–1926) and immediately recognized that the young man had the potential to become a brilliant engineer. Maybach became Daimler's protégé, and the two men would continue to work closely together for the remainder of the century.

A portrait of Gottlieb Daimler from 1895.

In 1869 Daimler brought Maybach with him when he was hired as director of the Maschinenbau Gesellschaft, a factory in Karlsruhe that manufactured locomotives. In 1872 Daimler became chief engineer of another factory, the Deutz Gasmotoren-

Fabrik, which produced various types of gas-powered engines, and he made Maybach head of design. At the Deutz factory the two men worked with Nicolaus Otto (one of the owners of the factory), who in 1876 developed the four-stroke engine known as the Otto Atmospheric Engine. Regarded as the best internal combustion engine of its time, the engine was still much too large for practical use in an automobile. Daimler envisioned a lighter engine, capable of reaching high speeds quickly.

NEW DESIGNS

After attempting to modify Otto's design, Daimler and Maybach left Deutz and set up their own manufacturing plant in Stuttgart in 1882. They were intent on developing light, fast-revving, gasoline-powered engines for use in motor vehicles. At first they encountered difficulties developing a reliable ignition system, but within a year Daimler had invented an ignition device with an incandescent tube in the cylinder head, which he patented in 1883.

Daimler and Maybach had also been occupied with a design for a smaller engine that could be used to power a moving vehicle. Their first experiment in this area led to their creating the world's first motorcycle. In 1885 the two men fitted a small gas-powered engine onto a specially designed wooden bicycle, which Daimler's son Paul rode for some six miles on its first outing. The inventors regarded the motorcycle, however, as merely a test run for more ambitious motorized vehicles.

The following year they fitted a more powerful, one-cylinder, water-cooled engine to a conventional four-wheel carriage. Soon they were attempting to use their engines to power boats, streetcars, and even dirigibles. At the 1889 world's fair in Paris, they exhibited a two-cylinder

Daimler considered the motorcycle to be a minor invention; however, the motorcycle has become a hugely popular vehicle in its own right. Here, champion rider Valentio Rossi tests a new bike in Sepang, Malaysia, in 2004.

A car built for the sultan of Morocco by Daimler and Maybach, around 1889.

V engine mounted in a new four-wheel car frame. This engine contained a carburetor that could inject evaporated gasoline mixed with air for use as the engine's fuel, a four-speed gearbox, and a belt-driven device for turning the wheels. The new vehicle could travel at about eleven miles per hour.

In 1890 Daimler founded the Daimler Motoren-Gesellschaft (Daimler Motor Co.) in Stuttgart. After only one year, both he and Maybach left the company because of disagreements with some of its leading investors, who had begun to take control of its direction. The two men continued to work independently on design ideas for automobiles (the other investors wanted the company to focus on the production of stationary engines). Daimler's cars made impressive showings in some of the early road races in Europe—including a victory in the first international auto race, the 1894 Paris-to-Rouen. Such successes allowed Daimler to resume control of the Daimler Motor Company in 1895. He once again made Maybach his chief engineer.

THE MERCEDES

This period saw the development of the car for which Daimler and Maybach would become best known—although Daimler himself would not live to see its introduction to the public. In 1900 Maybach, along

TIME LINE

1834	1853–1857	1863	1872	1882
Gottlieb Wilhelm Daimler born in Schorndorf, Württemberg, Germany.	Daimler works in a steam-engine locomotive factory in Strasbourg.	Daimler meets Wilhelm Maybach.	Daimler becomes chief engineer at Deutz Gasmotoren-Fabrik.	Daimler and Maybach leave Deutz to set up a manufacturing plant.

Jellinek, Mercedes, and Benz

Emil Jellinek enjoyed a life of high style on the French Riviera at the beginning of the 20th century. A wealthy diplomat for the Austro-Hungarian Empire who lived in Vienna, Austria, and Nice, France, Jellinek associated with many other wealthy people on the Riviera who wanted the finest, most advanced modern luxuries. An automobile enthusiast who had entered a number of road races in Europe (using his daughter's name, Mercedes, as his racing name), Jellinek knew he would be able to sell the best automobile available to his wealthy friends.

When he approached the Daimler Motor Company to produce such an automobile in 1900, Jellinek offered to purchase 36 cars on two conditions: first, he wanted to be the sole agent for the new car throughout the Austro-Hungarian Empire, as well as in France and the United States; second, he wanted the car to be named after his 11-year-old daughter, Mercedes. The German name "Daimler," he explained, could hurt the car's sales potential in France. Daimler's longtime partner Wilhelm Maybach, along with Daimler's son Paul, agreed to Jellinek's requests, and the Mercedes era was born. The name proved to be a great success, and soon all Daimler cars bore that name.

Karl Benz began Benz and Company in Mannheim, Germany, in 1883—a year after Gottfried Daimler started his first company in Stuttgart. Like Daimler, Benz was a brilliant engineer and automobile enthusiast who demanded the highest quality. The two companies produced cars and made various advances in automobile technology as competitors for some thirty years until, in 1926, they joined to form the Daimler-Benz Company. They marketed their cars under the trade name Mercedes-Benz. In the judgment of many, the Mercedes-Benz was the finest car made anywhere.

A display at the Mercedes-Benz Museum in Stuttgart, Germany, which opened in 2006.

TIME LINE (continued)

1883	1885	1890	1900
Daimler invents and patents an ignition device.	Daimler and Maybach begin testing small engines on various vehicles.	Daimler founds Daimler Motoren-Gesellschaft.	Daimler dies; Maybach and Daimler's son develop the Mercedes.

with Daimler's son Paul, developed what would be named the "Mercedes," after the daughter of Emil Jellinek, an Austrian diplomat living in France. Jellinek commissioned the Daimler Motor Company to produce a more powerful, front-mounted engine and a revolutionary body style consisting of a pressed-steel frame featuring an enlarged wheelbase and a lower center of gravity than previous frames. In return for the promise to purchase 36 cars, Jellinek insisted that the car be called Mercedes (see box, Jellinek, Mercedes, and Benz).

Gottfried Daimler died of heart disease on March 6, 1900. Although he did not live to see the groundbreaking design for the Mercedes, Daimler's name became forever associated with the car that gained worldwide recognition for its extremely high quality. The Mercedes became so successful that the Daimler Motor Company in 1902 decided to use the name for all its cars, taking out a trademark for that purpose. The Daimler Company merged with its longtime competitor the Benz Company in 1926, and the Mercedes-Benz was established as arguably the finest car in the world. Daimler is a pivotal figure in the transition from the "horseless carriage" to an automobile that would set standards of quality for decades.

—Paul Schellinger

Further Reading

Bird, Anthony. *Gottlieb Daimler: Inventor of the Motor Engine.* London: Weidenfeld & Nicolson, 1962.

Butterfield, Leslie. *Enduring Passion: The Story of the Mercedes Benz Brand.* Hoboken, NJ: Wiley, 2005.

Diesel, Eugen, et al. *From Engines to Autos: Five Pioneers in Engine Development.* Chicago: Henry Regnery, 1960.

See also: Ford, Henry; Honda, Soichiro; Transportation.

RAYMOND DAMADIAN

Inventor of the magnetic
resonance imaging scanner
1936–

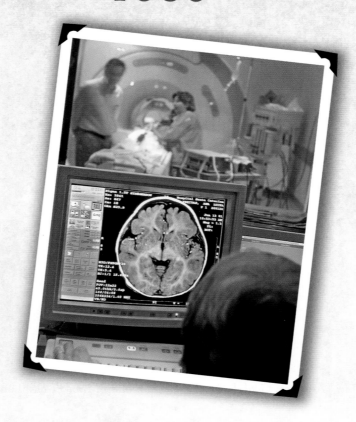

Raymond Damadian invented a device that has allowed medical practitioners to examine the interior of the human body with far greater clarity and
accuracy than any previous technology. The magnetic resonance imaging
(MRI) scanner provides much more detail than x-ray or CT scanner devices
and has truly revolutionized the field of diagnostic medicine.

Raymond Vahan Damadian was born on March 16, 1936, in Forest Hills, New York. As a child growing up in New York City, he entered the Juilliard School of Music Pre-College Program, where he studied violin for eight years. At age 15, Damadian began his university training at the University of Wisconsin on a Ford Foundation scholarship. While completing a degree in mathematics, he determined that medicine would offer more interesting career options. After graduating in 1956, Damadian entered the Albert Einstein College of Medicine in New York, receiving his medical degree in 1960.

While completing his medical degree at Einstein, Damadian made a crucial decision. He had entered medical school intending to specialize in internal medicine with a view toward becoming a clinical practitioner. He changed his mind, however, deciding to concentrate on medical research rather than on practicing medicine. One of his motivations came from watching his grandmother die of breast cancer. Damadian determined to make a difference in the way that cancer might be detected and thereby, perhaps, cured.

Upon receiving his M.D., Damadian served as a fellow in nephrology (the study of kidneys) at Washington University School of Medicine in

Raymond Damadian conducting an experiment in 1970.

A nuclear magnetic resonance spectrometer from 1965.

St. Louis and then as a fellow in biophysics at Harvard University's School of Medicine. At Harvard, he conducted advanced research in physics, physical chemistry, and mathematics. He went on to receive further training in the field of physiological chemistry at the School of Aerospace Medicine in San Antonio, Texas, while serving as a captain in the U.S. Air Force. Upon returning from active duty in 1967, Damadian accepted a faculty position in internal medicine and biophysics at the State University of New York (SUNY) Medical School Downstate.

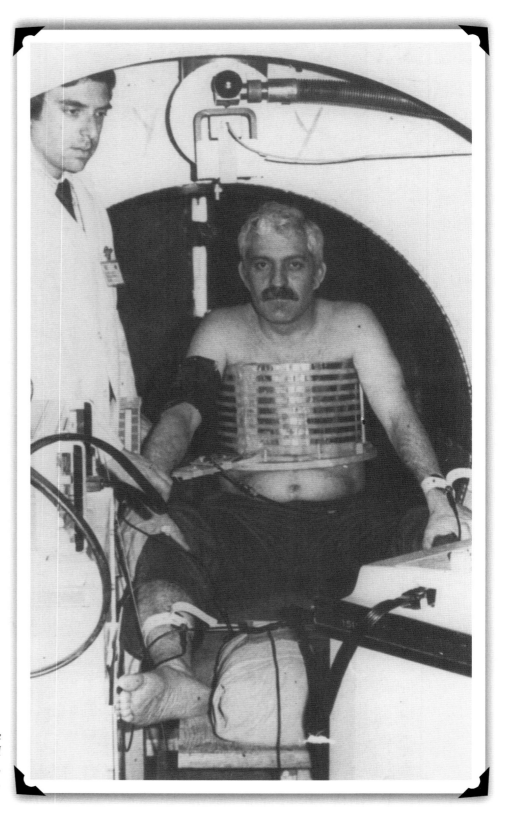

Damadian inside the first MRI machine, Indomitable.

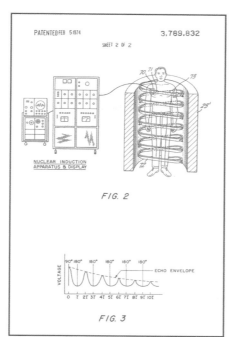

MAGNETIC RESONANCE SCANNING TECHNOLOGY

As a professor at SUNY, Damadian began working with a technology called nuclear magnetic resonance (NMR) spectroscopy. The NMR process involved placing an atom in a stable magnetic field, where its nucleus would absorb electromagnetic waves of radio frequency. Different types of nuclei would absorb these waves differently. Each nucleus would then reemit radio waves at a unique, predictable frequency, allowing researchers to analyze the composition of various substances with greater accuracy.

Isidor Isaac Rabi, a physicist at Columbia University, first discovered the NMR method in 1938; he was awarded a Nobel Prize in 1944 for his findings. Soon thereafter, physicists Felix Bloch of Stanford University and Edward Purcell at Harvard extended the method to analyze protons, part of the nucleus of an atom. Within a decade of Rabi's first discovery, NMR had become a standard tool for physicists and chemists to analyze substances. Although NMR spectroscopy had been used for identifying a wide range of substances, it had not been considered for use on humans.

Damadian first experimented with this technology while conducting research into sodium and potassium in living cells. He quickly saw the promise that NMR technology held out for diagnostic medicine. Specifically, he began to investigate how NMR could reveal the changes and differences between healthy cells and cancerous cells. In 1971, while experimenting with mice, Damadian discovered that NMR signals produced by cancerous cells lasted much longer than signals emitted by healthy cells. He published his findings in the journal *Science* and suggested that magnetic resonance scanning could be used to diagnose cancer early enough to slow or halt the disease's spread. He soon applied for a patent for a device that would allow doctors to use magnetic fields and radio waves to examine the human body for cancerous tumors. Damadian was awarded this patent in 1974.

Not everyone shared Damadian's enthusiasm for a magnetic resonance scanner for humans. After failing to get funding through the

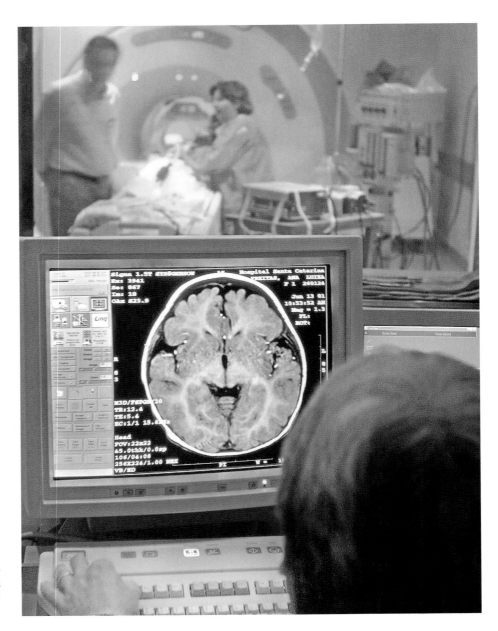

A doctor in Brazil examines a patient's MRI scan in 2001.

How Does MRI Work?

How did nuclear magnetic resonance become a way for creating images that doctors could examine? The answer lies in water.

The human body is about two-thirds water. Each water molecule is made up of hydrogen and oxygen atoms. When exposed to a magnetic field, the nuclei of hydrogen atoms react by changing their position slightly. As the magnetic waves pulse, the hydrogen atoms indicate specific differences in their nuclei as they react to the magnetic field and then return to their "normal" state. These differences (or oscillations) are detected through the resonance waves the nuclei emit.

When pathological changes occur inside our bodies (when we become sick), the water content of our organs and tissues changes. Raymond Damadian sought a way to show these changes in a three-dimensional image that would demonstrate the chemical structure of organs or tissues. Although he could analyze these data accurately, his method for processing the data into three-dimensional images of the chemical structure of tissue and organs was less successful. Other researchers, especially Paul Lauterbur, drew on Damadian's discoveries to create superior imaging technology.

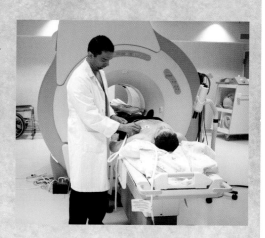

A doctor prepares his patient for an MRI scan.

National Institutes of Health and other government agencies, he finally managed to secure some funds from private backers to keep his NMR work going. After years of work, Damadian and his students unveiled their first machine, called Indomitable, which was capable of doing a magnetic resonance scan of the chest. Indomitable received wider attention in July 1977, after one of Damadian's students sat inside it for several hours while rudimentary images showed the student's heart and chest cavity.

In 1978 Damadian formed a company, FONAR (from Field fOcused Nuclear mAgnetic Resonance), to build MRI scanners. The first commercial scanner was released in 1980. The Food and Drug Administration

TIME LINE

1936	1960	1967	1971	1977	1978	1980	2003
Raymond Vahan Damadian born in Forest Hills, New York.	Damadian graduates from Albert Einstein College of Medicine.	Damadian joins the faculty at SUNY Medical School.	Damadian publishes his research about NMR signals.	The first NMR images are taken.	Damadian forms FONAR Company to build MRI scanners.	The first commercial scanner is released.	Approximately 75 million MRI scans are performed in one year.

approved Damadian's machine in 1984. Since that time, advances by Damadian and others have led to increasingly refined scanning technology that allows doctors to examine the human body in great detail.

CONTROVERSY OVER INVENTION

Damadian's first devices, built on the model of Indomitable, failed to sell because they produced low-quality images. While Damadian was developing his MRI scanner in the early 1970s, another medical researcher, Paul Lauterbur, at the State University of New York at Stony Brook, began to look at the results of Damadian's work and concluded that his method of generating images was flawed. Lauterbur proposed an alternative method using a second, weaker magnetic field that could be controlled to vary its position in relation to the first magnetic field, creating a two-dimensional image. Lauterbur's results were superior to Damadian's, and eventually Damadian himself adopted Lauterbur's methods for his production of MRI scanners at FONAR.

The two men never collaborated in their research and were, in fact, often at odds with each other. Damadian felt that Lauterbur had unfairly ignored his contributions, as Lauterbur attempted to claim the development of MRI scanning technology for himself. In a paper he published in the journal *Nature* in March 1973, Lauterbur demonstrated images achieved by his magnetic resonance scanning without any reference to Damadian or his work. This began an adversarial relationship between the two men that culminated in a controversial awarding of the 2004 Nobel Prize in Physiology or Medicine to Lauterbur and fellow British researcher Peter Mansfield for development of the MRI. Although the Nobel committee could have awarded the prize to three researchers, Damadian was omitted. He subsequently took out several full-page ads in newspapers protesting his exclusion and urging readers to express outrage to the Nobel committee.

Some maintain that Damadian is the true inventor of MRI because he was the first to adapt known NMR principles to medicine. Supporters of Damadian also point out that he was the first to recognize that cancerous cells emit signals of longer duration than healthy cells when scanned. Others, recognizing the flaws in Damadian's images (the crucial "I" of MRI), consider that the most important developments in MRI research rightfully belong to Lauterbur and Mansfield. Another factor often cited as a reason for Damadian's exclusion from the Nobel Prize is his religious philosophy. As a creationist (one who does not believe in evolution), Damadian is part of a minority in the scientific world. Some creationists have decried the Nobel snub as prejudice against their beliefs.

Damadian was awarded the National Medal of Technology by President Ronald Reagan in 1988.

Damadian has been recognized elsewhere for his contributions to MRI technology. He was awarded, jointly with Lauterbur, the National Medal of Technology by President Ronald Reagan in 1988, "for their independent contributions in conceiving and developing the application of magnetic resonance technology to medical uses including whole-body scanning and diagnostic imaging." Damadian was inducted into the National

Inventors Hall of Fame in 1989, and in 2001 he received the Lemelson-MIT Lifetime Achievement Award. The first scanning machine Damadian built is part of the Smithsonian Institute's permanent collection

Magnetic resonance imaging is used today to detect everything from torn ligaments (athletic facilities are increasingly equipped with MRI scanners) to abnormalities in human tissues and cells. In 2003, approximately ten thousand MRI machines were in use throughout the world. That year, some 75 million MRI scans were performed. The technology Raymond Damadian helped to create has become an integral part of medical diagnosis in the 21st century.

—Paul Schellinger

Further Reading

Books

Hashemi, Ray H. *MRI: The Basics*. London: Lippincott Williams and Wilkins, 2003.

Mattson, James, and Merrill Simon. *The Pioneers of NMR and Magnetic Resonance in Medicine: The Story of MRI*. Ramat Gan: Israel Bar-Ilan University Press; Jericho, NY: Dean, 1996.

Web sites

FONAR
> The home page of Damadian's company.
> http://www.fonar.com/

Lemelson-MIT Program: Raymond Damadian
> Features a biography of Damadian and a link to a brief video.
> http://web.mit.edu/invent/a-winners/a-damadian.html

Nobel Prizefight
> Further information on the controversy over the 2003 Nobel Prize.
> http://whyfiles.org/188nobel_mri/

See also: Cormack, Allan, and Godfrey Hounsfield; Health and Medicine.

JOHN DEERE

Inventor of improved steel plow
1804–1886

Large-scale agriculture in America, especially in the grain-growing regions of the Midwest, became possible as a result of John Deere's invention of an improved plow. The plows that settlers brought with them as they migrated from the East were no match for the hard clay soil they encountered in the West. Deere's invention made farmers realize that they could make a living out of this fertile, but often nearly untillable, soil.

EARLY YEARS

John Deere was born February 7, 1804, in Rutland, Vermont. His father, William Rinold Deere, had emigrated from England; his mother, Sarah Yates, was born in Connecticut. When John was four years old, his father was lost at sea. His mother was left to raise John and his five siblings in circumstances verging on poverty.

Deere received only a basic education, attending local schools while working to assist his family financially. At age 17 Deere began an apprenticeship to a blacksmith. He spent four years as apprentice to Captain Benjamin Lawrence in Middlebury, Vermont, before setting up shop as an independent blacksmith. Deere, who worked at this trade for the next 12 years at various locations throughout Vermont, made a good name for himself, as he was devoted to the highest-quality standards in whatever he worked on—from shoeing horses to producing household and farm implements ranging from skillets to hay rakes.

MOVING WEST

Deere married Demarius Lamb in 1827. The family moved around in central Vermont as John sought steady work and their financial circum-

An undated engraving of John Deere.

stances remained difficult. In 1837, with four children and another on the way, the 33-year-old Deere decided to move West, to Illinois. He planned to establish some kind of business in the newly opened western regions of Ohio, Indiana, and Illinois, and then to send for his family once he was established. He sold his blacksmith shop in Vermont and, with about $70 in his pocket, began the journey to Grand Detour, Illinois— a location on the Rock River where a former employer of Deere's had settled. While back in

Vermont to retrieve his own family, this man had told Deere that the prairie offered great opportunity.

After traveling several weeks by boat and wagon through canals and over rough land, Deere arrived in Grand Detour and quickly set up a small blacksmith shop on a rented plot. Soon he was getting work from farmers and other settlers in the area. Many of these men hailed originally from Deere's native New England, where they had learned to till the relatively light, sandy soil characteristic of that region. Out on the western prairie, the tools—particularly the plows—they had brought with them proved unequal to the very different soil they found there. These plows, made of cast iron and constructed at a poor angle for handling heavy soils, easily became clogged. Farmers complained that they had to stop every few feet to knock clods of sticky soil off their plows by hand. After hearing many such complaints, John Deere began to work on making a plow more suitable for the land.

> I will never put my name on a product that does not have in it the best that is in me.
>
> —John Deere

THE SINGING PLOW

In 1837, still in his first year in Grand Detour, Deere came up with a solution—using a broken steel saw blade, Deere fashioned a plow that effectively cut through the soil without becoming clogged. Its highly polished surface was self-cleaning, and the plow itself literally sang through the earth: Deere's plow would eventually become known as the "singing plow" because of the unique, high-pitched sound it made as it cut neatly through the soil.

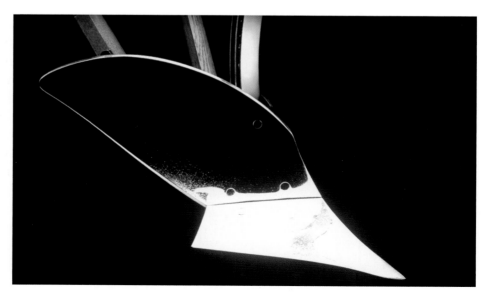

The steel blade of John Deere's "singing plow," designed in 1837.

His family joined him in Illinois, and Deere began to build his reputation as a plow maker, even though he could produce only a few plows a year at first because of his limited resources. Having introduced the steel plow, Deere was always concerned that someone else would come along and produce a better one. Thus he was constantly trying to improve his design—sometimes over the objections of his employees, who complained that changing the design so frequently made their work more difficult. Yet Deere insisted on making improvements whenever he had the means to do so.

The kind of steel Deere wanted could be found only in England. Transported across the Atlantic and then shipped up the Mississippi River to the Illinois River, these special rolls of steel were costly. In response, Deere made two important moves. First, in 1848, he moved his business and his family to Moline, Illinois, on the Mississippi River, in order to take advantage of the improved transportation and water power that the great river offered. Second, he negotiated with a steel mill in Pittsburgh, Pennsylvania, to produce the specific steel he required. Soon he was manufacturing more than 1,500 plows a year.

PRODUCTION EXPANDS

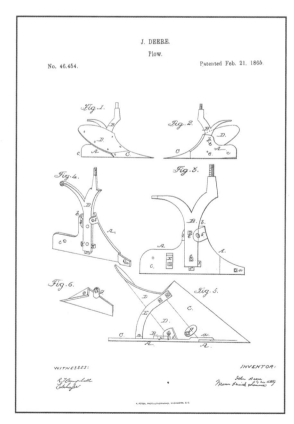

Illustrations from John Deere's patent for his plow improvements. Figures 1 and 2 are side elevations of the plow; figures 3 and 4 are different side elevations that show the inside of the plow; figure 5 is a bottom view; figure 6 is a perspective view.

As more and more farmers heard of Deere's singing plow, demand increased. Plows were shipped by boat up and down the Mississippi River and by wagon into the countryside. Many of them were taken west by settlers hoping to tame their portion of the vast territory beyond the Mississippi. By 1857, Deere's company was producing 10,000 plows a year.

John Deere's oldest surviving son, Charles, began working at the company in 1853. In the wake of the nationwide financial crisis of 1857, John Deere incorporated

The John Deere Tractor

Although his name is synonymous with farm and lawn tractors, John Deere never even imagined such a technology; nor did he invent it. Neither did his son Charles, who guided the company from around the middle of the 19th century into the early years of the 20th. Under Charles, Deere and Company expanded its product line from the steel plow to a wide range of farm implements—cultivators, planters, and many others—but it still had no connection with tractors.

In fact, Deere and Company entered the tractor business only in 1918, when, under the company's third president, William Butterworth, it acquired the Waterloo Gasoline Traction Engine Company of Waterloo, Iowa—producers of the Waterloo Boy tractor. This move, coming after several other acquisitions of farm equipment companies beginning in 1911, reinforced Deere and Company's position as a dominant American farm manufacturer. After continuing to produce and market the Waterloo Boy for several years, in 1923 Deere and Company introduced its first tractor bearing the John Deere name: the classic Model D would be produced until 1953.

A John Deere Model G in an antique tractor pull, held in Oklahoma in 1999.

the company and brought his son into partnership. Having worked in positions from bookkeeper through various marketing and sales posts, Charles Deere proved to be an extremely capable businessman. The company was reincorporated as Deere and Company in 1868. Although John Deere would remain president of the company until his death in 1886, his son Charles ran the day-to-day operations from 1858 onward.

After 1858, John Deere's involvement in the company he had founded focused on product development and sales. He also began to give more of his time to social and philanthropic causes, eventually entering local politics. He served in a variety of civic roles in Moline before becoming mayor in 1873. Until his death there on May 17, 1886, at the age of 82,

TIME LINE

1804	1825	1837	1857	1858	1873	1886
Deere born in Rutland, Vermont.	Deere sets up shop as a blacksmith.	Deere invents the "singing plow."	Deere's company produces 10,000 plows per year.	Deere's son, Charles, takes over the day-to-day operations of the company.	Deere becomes mayor of Moline, Illinois.	Deere dies.

John Deere contributed generously to local institutions and causes; he was widely mourned at his death.

DEERE AND THE AMERICAN WEST

John Deere was one of a few people who truly opened up the West to expansion. His singing plow was an indispensable tool for farmers in the untamed new lands.

The company continues to operate under the name Deere and Company, with headquarters still in Moline. Today Deere and Company is best known for its tractors and lawn mowers. Important as these technologies have been in American society, they have not revolutionized a nation or affected the movement of populations as profoundly as the steel plow invented by the company's founder in 1837.

—Paul Schellinger

Further Reading

Books

Broehl, Wayne, Jr. *John Deere's Company: A History of Deere and Company and Its Times*. New York: Doubleday, 1984.

Clark, Neil M. *John Deere: He Gave the World the Steel Plow*. Moline, IL: Deere, 1937.

Web site

Deere and Company
   Detailed biography and chronology.
   http://www.deere.com/en_US/compinfo/history/johndeere.html

See also: Colt, Samuel; Food and Agriculture; McCormick, Cyrus Hall.

GEORGE DE MESTRAL

Inventor of Velcro

1907–1990

The invention of the Velcro hook-and-loop fastener is considered one of history's "happy accidents." Swiss engineer George de Mestral did not set out to invent a new type of fastener in the 1940s, but, inspired by something he discovered after a hike in the mountains, that is exactly what he did. Patented in 1955, Velcro fasteners have since become ubiquitous, used everywhere from the runway to the space shuttle.

EARLY YEARS

De Mestral was born in 1907, in a small town outside of Lausanne, Switzerland, not far from the country's capital, Geneva. He exhibited an inventor's spirit early—by age 12, he had already designed and patented a toy airplane. He attended the Swiss Federal Institute of Technology (École Polytechnique Fédérale de Lausanne), one of the premier universities in Europe, and earned a degree in electrical engineering.

After graduation, de Mestral began work as an engineer, starting in the machine shop of a local firm. His love for the outdoors led to his becoming an amateur mountaineer.

A WALK TO REMEMBER

During one excursion in the Swiss mountains in 1941, de Mestral took special note of the burrs that had become attached to his wool pants and tangled in his dog's fur. The burrs were relatively small in size, but clung tenaciously to clothing and fur—so tenaciously, in fact, that they sometimes had to be cut off.

Intrigued by the burrs' strength, de Mestral examined one under a microscope. What he saw was a maze of hundreds of tiny hooks that snared the woven threads in cloth or strands of animal hair. One tiny hook seemed insignificant and had little strength, but hundreds of hooks together were quite powerful.

De Mestral's inspiration for his most famous invention came to him in the Swiss mountains.

A close-up of hook-and-loop fasteners shows how the rough side, which is composed of tiny hooks, grabs onto the soft side, which is composed of tiny loops, to create a very strong fabric fastener.

Others may have written off the burrs as a nuisance, but de Mestral saw them as an opportunity. He believed he could invent a new type of practical fastener based on the hook-and-loop principle of the burr.

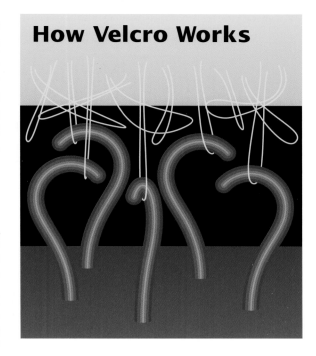

How Velcro Works

THE PATH TO PERFECTION

Despite the simplicity of the burr's design, de Mestral needed nearly ten years to create Velcro. Frustrated with his initial attempts to create a type of hook-and-loop "tape" using the burr principle, de Mestral sought out the expertise of a French fabric weaver and a Swiss loom maker.

The first prototypes were made of cotton, which did not offer the kind of strength de Mestral had marveled at with the burrs. De Mestral then discovered nylon.

Nylon, a revolutionary type of synthetic fiber invented by Wallace Carothers (1896–1937) in the 1930s and dubbed the miracle fiber, was well known for its strength and durability. While experimenting with this new fiber, de Mestral discovered that when nylon was sewn under infrared light, it hardened, and when the fibers were cut, they formed nearly indestructible hooks.

Even with the new nylon technology, perfecting his invention took time. At first, the loops were the wrong size for the hooks, or vice versa; finding the perfect angle at which to cut the nylon fabric was a time-consuming challenge. Finally, Velcro was born.

De Mestral applied for a patent from the Swiss government in 1951. His prototype consisted of two strips of nylon with roughly three hundred hooks per square inch on one side and three hundred loops on the other. De Mestral was granted U.S. Patent No. 2,717,437 in 1955.

PRODUCTION VALUES

De Mestral founded Velcro S.A. in Switzerland in 1952 and soon after opened a Velcro factory in Aubonne, Switzerland. By the late 1950s,

Made by Mother Nature

Velcro is revolutionary for reasons far beyond its strength and durability. It is often recalled as the first real instance of biomimetics—biological mimicry, or the practice of using forms found in nature to inspire engineering or design.

The burrs that inspired de Mestral had evolved as a means of pollination—the hooks ensured that the plant seeds contained within the burr could be spread far and wide. The burr design had secured the success of the plant species for eons. De Mestral recognized the potential of the hook-and-loop design and successfully duplicated it with Velcro. Little did he know that, decades later, an entire academic discipline would spring up around this very act.

De Mestral took his inspiration from burrs, which are crucial for pollination in some plant species.

de Mestral had been granted patents not just in the United States but also in other countries throughout Europe and Canada. Velcro factories were established throughout the Western Hemisphere. The U.S. factory, which remains in operation to this day, began producing Velcro in Manchester, New Hampshire, in 1958. (With such a long manufacturing history, the Velcro factory workers in Manchester have earned the nickname "Velcroids.")

De Mestral believed Velcro would revolutionize the garment industry, replacing standard clothing fasteners such as zippers and buttons. However, his invention was poorly received at first. The garment industry shunned Velcro, claiming it was too unsightly to use on clothing. For many years, de Mestral earned an average of only $55 per week on his product. Then the burgeoning aerospace industry discovered Velcro. Aircraft engineers began using it to fasten or secure various parts of their planes, from insulation to seat panels. At the end of the 1950s, more than sixty million yards of Velcro were being produced annually.

Still, Velcro's big break did not come until the 1960s, when the National Aeronautics and Space Administration (NASA) began sending

Hook-and-loop fasteners have found many uses in the modern world, particularly in clothing and devices like arm and leg braces.

rockets into outer space. NASA engineers used Velcro to secure items in zero-gravity environment. Images of space travel caught the public imagination and introduced Velcro to mainstream America. Indeed, many people mistakenly believed Velcro had been designed by NASA.

INVENTOR FOR LIFE

Even after selling his Swiss company, de Mestral continued to earn royalties from his invention, estimated to be in the millions. He also continued to invent—a pillbox designed to help organize daily medications, a device for measuring humidity, a commercially successful asparagus peeler. Mainly, de Mestral dedicated himself to inspiring and supporting other, often younger, inventors. He died on February 8, 1990, in Commugny, Switzerland.

Almost a decade after his death, de Mestral was inducted into the Inventors Hall of Fame for creating Velcro. At the event in 1999 he was

TIME LINE

1907	1941	1952	1958	1960s	1978	1990	1999
George de Mestral born in Lausanne, Switzerland.	De Mestral is inspired to create hook-and-loop fasteners during a walk in the mountains.	De Mestral founds Velcro S.A.	The U.S. branch of de Mestral's company opens in Manchester, New Hampshire.	NASA's use of hook-and-loop fasteners vastly increases the popularity of de Mestral's invention.	De Mestral's patent expires, opening the door to competitors.	De Mestral dies in Switzerland.	De Mestral is posthumously inducted into the Inventors Hall of Fame.

represented by his son, who said, "He believed that what mattered most was not the money he received for his work, but the esteem he received from his contemporaries."

VELCRO NOW

"Velcro" is the short name of a set of companies formally called Velcro Industries B.V. The Velcro companies have become synonymous with the product they sell; however, the product itself is still officially known as the "hook-and-loop fastener." Like Kleenex in the facial tissue market or Xerox in the copier market, Velcro the brand name has come to define an entire industry. However, the company is not without competition. Copycat products sprang up even before the Velcro patent expired in 1978. Now, Velcro faces stiff competition from Japanese manufacturers YKK and Kuraray and others.

Technical advances made Velcro acceptable to the garment industry, and it can now be found on everything from children's shoes and high-tech sporting equipment to high fashion. It is still used by NASA and the aerospace and automobile industries; it has even found use in the world of medicine, where it helped hold together a human heart during the first artificial heart surgery. A small-scale version of Velcro can be used to close wounds. Velcro has plenty of entertaining uses as well—talk-show host David Letterman famously invented Velcro jumping, in which a person wears Velcro-hook clothing and tries to jump and stick to a Velcro-loop wall.

The word Velcro derives from the French words *velours* (velvet) and *crochet* (hook).

—Laura Lambert

Further Reading

Articles

Quinn, Jim. "Burrs Inspire Man to Create New Fastener." *Akron Beacon Journal,* September 13, 1999.

Singh, Simon. "Serendipity a Bit of a Snag." *The Independent* (London), June 13, 1999.

See also: Carothers, Wallace; Cloth and Apparel.

RUDOLF DIESEL

Inventor of the diesel engine
1858–1913

Trucks, trains, ships, submarines, and power plants all owe a debt to the inventiveness of German engineer Rudolf Diesel. In the late 19th century, Diesel came up with an idea for an improved mechanical engine that could produce more power from less fuel. This idea promised to make him rich, but Diesel died unexpectedly, leaving other inventors to develop his ideas into engines that power the world today.

EARLY YEARS

Rudolf Diesel's parents were German and he lived much of his life in Germany, but he was born in Paris, France, on March 18, 1858, and spent his childhood in that city. When Diesel was 12, in 1870, war broke out between France and Germany, making it unsafe for Germans to remain in France. The family fled to England. Rudolf was soon sent to Germany, to his father's hometown, Augsburg, to continue his schooling. Later, he went to the Technische Hochschule (Technical High School) in Munich to study engineering.

There Diesel became the pupil of Carl von Linde (1842–1934), a brilliant young German engineer who was in the process of inventing the technology used in modern refrigerators. By all accounts, Diesel was a dedicated student. In von Linde's laboratory, he learned about thermodynamics (the science of how heat moves) and heat engines (machines that extract the power locked in fuels such as coal and gasoline by alternately heating and cooling such fuels). In 1878, von Linde gave up his lecturing position to start a company manufacturing refrigerators. Two years later, Diesel joined him.

Rudolph Diesel, photographed in 1912.

BETTER THAN STEAM

The 19th century was the heyday of the steam engine, which an Englishman, Thomas Newcomen (1664–1729), had invented around 1712. A steam engine is a type of heat engine that works like a giant kettle, burning coal to boil water and using the resulting steam to drive railroad locomotives and other machines. In a steam engine, the coal is burned to produce steam in one place (the boiler) and the steam makes useful power in another place (the cylinder, a large metal can with a piston that slides back and forth inside it). This process is called external combustion, because the combustion (burning of the fuel) happens outside the cylinder. As the coal is burned, however, and the heat is transferred from the boiler to the cylinder, a great deal of energy is wasted, making steam engines very inefficient.

> The automobile engine will come, and then I will consider my life's work complete.
>
> —Rudolf Diesel

During the 19th century, a number of European inventors tried to develop engines that would use internal combustion. Instead of burning fuel outside the cylinder, the idea was to burn it inside the cylinder—a more efficient process that would waste less energy. The first person to suggest this, around 1800, seems to have been a French inventor named Philippe Lebon (1769–1804). Over the following decades, many other Europeans tried to make engines of this kind, with little success.

Finally, in 1876, German engineer Nicolaus Otto (1832–1891) made the first successful internal combustion engine. Otto's invention was a great improvement on the steam engine. It burned fuel and air inside a large metal cylinder in four separate stages or "strokes," known as the Otto cycle. In the first step, air and fuel entered the cylinder and mixed together. In the second step, a piston moved into the cylinder and compressed (squeezed) the mixture. This prepared for the third step: a spark of electricity set the mixture alight, causing a miniature explosion that pushed the piston back out of the cylinder, turned a crank, and generated power. In the fourth step, the piston pushed back into the cylinder to remove the exhaust gases produced by the burning fuel. This four-step cycle repeated itself, turning a constant supply of fuel into mechanical power. Otto's engine was ultimately reworked and revised into the gasoline engine that still powers most cars today.

INVENTING THE DIESEL ENGINE

Diesel was unimpressed by Otto's engine. In his studies of thermodynamics, Diesel had learned that the best possible heat engine follows a series of steps called a Carnot cycle, named for its discoverer, French physicist Nicholas Sadi Carnot (1796–1832). Although Otto's engine

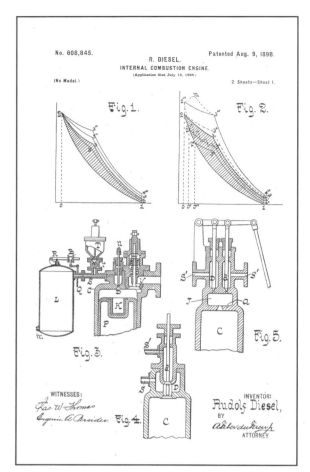

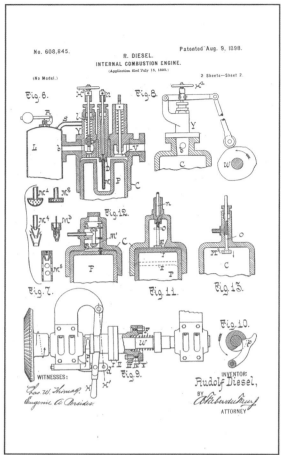

was more efficient than a steam engine, it was less efficient than it might have been. Diesel believed he was a scientific genius and was certain he could make a better engine—and this was one reason he became an inventor.

The other reason was more ambitious. The Industrial Revolution, driven largely by the steam engine, had transformed society. Steam engines had made a small number of industrialists very wealthy and powerful, but ordinary people saw less benefit. Many struggled in dirty and unhealthy factories operating the machines; others worked long hours in underground mines digging the coal to power steam engines; and all benefited financially much less than their bosses. Rudolf Diesel realized that engines put power into the hands of ordinary people. He imagined a smaller, less expensive engine, running on affordable fuel, that would help artisans and small industrialists compete with the wealthy and powerful. Diesel's engine would be the people's engine.

Around 1890, Diesel figured out how to make his better internal combustion engine. Instead of using an electric spark, he burned the fuel-air mixture another way. When gases are compressed, they can heat up greatly—this is why a bicycle pump often gets hot during use. Diesel used this effect to burn fuel in his engine. He built the engine around a long (10-foot) sturdy metal cylinder with a free-moving piston inside it. First, the cylinder filled with air. Then the piston compressed the air to about 1/20th of its original volume, making it extremely hot. Next, a pipe squirted droplets of fuel into the cylinder. Because the air and the cylinder were so hot, the fuel caught fire at once and gave off exhaust gas, which pushed the piston back out of the cylinder and turned a crank, much as in an Otto engine.

By 1892, Diesel had figured out the theory of his engine completely and was granted a patent. He published his ideas in 1893, and on August 10 of that year, his engine first ran successfully under its own power. The engine was still experimental and Diesel spent several more years improving it. In December 1896, he demonstrated an engine that was about twice as efficient as a steam engine, getting twice as much power from the same amount of fuel by wasting less energy.

Truckers protest rising prices of diesel fuel in front of the Capitol Building in 2000.

DEATH OF AN ENGINE MAKER

From 1897 onward, Diesel promoted the invention tirelessly—so much so that he became ill from nervous exhaustion. His arrogance did not help. He claimed that few of the world's factories were good enough to build his engine and most should not even bother to try. In 1899, he set up his own plant in Augsburg to manufacture the device. He was no businessman, however, and work proceeded slowly because he had not ironed out all the problems associated with his engine's manufacture. Nevertheless, other engineers were impressed by what he had done and bought the rights to manufacture diesel engines to his design. This guaranteed he would soon become very wealthy.

Then tragedy struck. On September 29, 1913, while Diesel was traveling to London from Germany on the mail ship *Dresden*, he mysteriously vanished and was assumed to have fallen overboard. His body was found by another boat some days later, identified from papers in his pockets, and then buried at sea. Diesel's unusual death immediately aroused suspicion. Some people believe he committed suicide: he had always been a sensitive man, and the strain of working too hard may have caused him to take his life. Others have suggested he was mur-

In 1897, Diesel (center), his wife, and their children greet visitors at their Munich home.

Inventing Biodiesel

The gasoline and diesel burned in most vehicle engines comes from petroleum, a thick, brown liquid formed when plants and animals decay beneath rocks, either on land or off-shore. Over millions of years, high temperatures and pressure turn the once-organic remains into a mixture of many different hydrocarbons (complex chemicals formed from the elements hydrogen and carbon). When these fossil fuels are burned in air, a chemical reaction called combustion takes place. This unlocks the energy stored in the fuel, producing heat that can power an engine. However, most materials will burn in air, so, theoretically, fuels other than petroleum could be used to drive engines.

Most diesel engines burn a low-grade and less expensive product of petroleum known as diesel—but that was never Rudolf Diesel's intention. One of his dreams was to create an engine that could run on almost any kind of fuel, even oil made from vegetables. He believed that this versatility would bring many benefits to society. It would free people from the need to buy expensive petroleum-derived fuel. Also, if communities grew crops to power their engines, farmers would benefit. Diesel's vision was that his engine could change society by putting more power into the hands of ordinary people and local communities.

In the 1890s, the idea of making an engine that people could power with plant products seemed outrageous; today, when demand for oil is high and people have greater concern for the environment, it seems less so. Unlike petroleum, which is a finite resource, oils made from animal or vegetable fats are renewable: they could last forever. Many people are already exploring these alternative fuels. In the United States, an estimated two million people drive dual-fuel cars, which can run on either alcohol derived from plants or traditional gasoline at the flick of a switch.

Elsewhere in the world, interest is growing in biodiesel: a fuel that can be made from many different organic materials, including soybeans, palm oil, peanut oil, hemp, animal and vegetable fats—and even the waste grease from cooking. Unlike petroleum, biodiesel has many environmental benefits. When it is produced from growing crops, it does not add to the problem of global warming. It is also nontoxic and produces fewer harmful emissions than traditional diesel or gasoline. Some states are already actively promoting biodiesel. In Minnesota, for example, a state mandate has required all diesel fuel to contain at least 2 percent biodiesel since fall 2005. The famous country singer Willie Nelson has even launched his own brand of biodiesel, Bio-Willie, which he promotes as a homegrown product that can help family farmers.

dered by either German or French agents who wanted to stop him from selling the secrets of the diesel engine to the British. However, it is equally likely that he simply fell overboard and drowned.

THE POWER OF THE DIESEL ENGINE

The first commercial diesel engine was built not in Germany but in the United States. Adolphus Busch (1839–1913), the brewer of Budweiser beer, saw one of Diesel's models at an exhibition in Germany and realized how useful the invention would be. He bought the North American rights and built his own version, with Diesel's help, at his brewery in St. Louis, Missouri, in 1898. Working with Sulzer Brothers, a Swiss company, Busch developed an improved diesel engine and sold it throughout the United States and Canada. Busch-Sulzer engines were used in ships, ferryboats, factories, and power plants; during World War I (1914–1918), they powered many U.S. submarines.

Diesel trains in Los Angeles, California, in 2004.

Diesel was convinced he had developed a brilliant invention, but his engine was not an immediate commercial success. Only after his death, when it became easier for engineers to experiment with his design, were smaller, more practical, and more successful diesel engines

TIME LINE

1858	1890	1892	1896	1899	1913
Rudolf Diesel born in Paris, France.	Diesel devises his internal combustion engine.	Diesel receives a patent for his engine.	Diesel demonstrates an engine that is nearly twice as efficient as the steam engine.	Diesel establishes his own plant in Augsburg, Germany, to manufacture his engine.	Diesel dies mysteriously while en route to England.

produced. In the United Kingdom, Dugald Clerk (1854–1932) simplified Diesel's four-stroke engine so that it worked using only two stages instead of four. That made the engine smaller and simpler. Two-stroke diesel engines were used in smaller power plants, water pumps, and ships. Shortly afterward, a compact "semi-diesel" engine was developed that combined the best features of the Otto and Diesel engines. It worked at lower pressures than a Diesel engine and used an electric spark to ignite the fuel, just like an Otto engine. The diesel-electric engine, which was a diesel engine connected to an electricity generator, was another modification. As this engine spun around, it turned the electricity generator and produced electricity to drive an electric motor at a much lower speed.

The first diesel engines were so large and heavy that they had to be fixed in place and could be used only in large factories and power plants. The efforts of many different engineers led to diesel engines' gradually becoming small enough to be used in vehicles. They were used first to power large ships and railroad locomotives and later in smaller vehicles such as trucks and buses. The first diesel automobile was demonstrated in Germany in 1922, though diesel cars needed a sharp increase in gasoline prices, such as those of the 1970s, to become popular.

In one sense, nothing could be more mundane than the noisy, dirty diesel engines that drive the modern world. More than a century after they were invented, diesel engines power the majority of trucks, buses, locomotives, ships, and submarines on the planet. Their continuing popularity is attributable to their high efficiency (they consume less fuel) and the relatively low cost of diesel oil (a lower-grade by-product of petroleum, from which gasoline is also made). However, it is all too easy to take an everyday invention like the diesel engine for granted. Just as Rudolf Diesel orig-

Biodiesel enthusiasts honor Diesel as a visionary and a pioneer: each August 10—the day Rudolf Diesel first successfully ran his engine—they celebrate International Biodiesel Day.

inally hoped, his engine—and the gasoline engine developed by Nikolaus Otto—put greater power into the hands of ordinary people. Few could afford to buy or run an enormous steam engine, but many were soon able to run vehicles, machine shops, or factories using diesel engines. Mundane the diesel engine may be, but nothing could be more revolutionary.

—Chris Woodford

Further Reading

Books

Challoner, Jack. *Eyewitness: Energy.* New York: Dorling Kindersley, 2000.
Woodford, Chris. *Power and Energy.* New York: Facts On File, 2004.

Web sites

Diesel Technology Forum
    Information about the latest developments in diesel engines.
    http://www.dieselforum.org/
How does a car engine work?
    A video animation explaining the inner workings of a car engine.
    http://www.brainpop.com/technology/transportation/cars/
How does an internal combustion engine work?
    Animated artworks explaining the different stages of an engine's operation.
    http://www.keveney.com/otto.html
Willie Nelson's Biodiesel
    A comprehensive site about biodiesel from singer Willie Nelson.
    http://www.wnbiodiesel.com/

See also: Ford, Henry; Newcomen, Thomas; Otto, Nicolaus August; Transportation.

CHARLES DREW

Inventor of the blood bank
1904–1950

Some inventions make life more interesting or enjoyable; others make work quicker or easier. Medical inventions tend to have a more profound effect on people because they can literally make the difference between life and death. When African American doctor Charles Drew invented a way of storing blood—the blood bank—he helped to save millions of lives. Ironically, his invention was not enough to save his own life after he was involved in a tragic automobile accident.

EARLY YEARS

Charles Drew was born in Washington, D.C., on June 3, 1904, the oldest of the five children of Richard and Nora Drew. Both athletically gifted and very enterprising, Drew was frequently voted best athlete and earned four varsity letters at Dunbar High School. When he was not on the football field or the track, he was running a business with his friends, selling newspapers.

Drew's athletic gifts earned him a scholarship to Amherst College in 1922, but, coming from a poor family, he still had to wait tables to pay his way. For his first two years at Amherst, he excelled on the sports field: he was a champion quarterback and baseball player and was an outstanding hurdler. However, this early success came at the expense of his grades. Gradually, he focused more on his studies, managing to earn 100 percent on his final chemistry exam. When he graduated from Amherst in 1926, Drew was still not sure what he would do with his life. For the next two years, he worked as a science teacher and athletics coach at Morgan State College in Baltimore, Maryland. He and the school were successful—the college's basketball and football teams won championships.

When Drew decided to apply to medical school, the money he had saved at Morgan State proved useful. He attended McGill University in Montreal, Canada, where he studied with eminent British anatomy professor Dr. John Beattie. Drew's brilliance was soon recognized: when he graduated in 1933 with degrees in medicine and surgery, he ranked second of 127 students. Now a doctor, he worked as an intern in Montreal before moving back to the United States—to Howard University in Washington, D.C., in 1935.

STORING BLOOD

While working as an instructor in pathology at Howard, Drew won a Rockefeller Foundation Fellowship to study at Columbia University in New York. There, between 1938 and 1940, he began his important research into blood—and he soon became a world expert on transfusions (drawing blood from a healthy person to give to another who is

A portrait of Charles Drew by Betsy Graves Reyneau, about 1943.

How Blood Groups Work

Donor		Recipient			
		O	A	B	AB
People with blood type O can donate to all other groups (universal donor).	O	⬤	⬤	⬤	⬤
People with blood type A can donate to groups A and AB.	A		⬤		⬤
People with blood type B can donate to groups B and AB.	B			⬤	⬤
People with blood type AB can donate only to other ABs but can receive donations from all types (universal recipient).	AB				⬤

Drew was able to separate plasma from red blood cells, which come in different types.

sick or injured). In 1940, he published everything he had discovered in his doctoral dissertation, "Banked Blood: A Study in Blood Preservation."

Although blood transfusions were routine at the time, they were problematic. People have different types (or groups) of blood, and in most cases blood donated by a person of one group cannot be given to a person of a different group. In Drew's time, blood had a maximum storage life of one week and could be stored for only two days in a refrigerator before it began to break down. Thus, any stockpiling of blood for future emergencies was impossible.

During the course of his research, Drew realized that blood transfusions could be carried out in an entirely different way. Blood has two main components: the red blood cells that carry oxygen and other vital supplies to different parts of the body; and plasma, the pale, watery part of the blood, wherein the red blood cells are suspended. The red blood cells make one blood group different from another and one person's blood incompatible with another's; the red blood cells also break down fastest. However, everyone has the same type of blood plasma.

Charles Drew saw that it was relatively easy to divide blood into its different components and store them separately. When the plasma was

Drew (left) and members of a mobile Red Cross unit from Presbyterian Hospital, New York City, in 1941.

separated from the red blood cells, he found it could be stored under refrigeration much longer. He began to theorize that blood plasma could be used by itself for emergency transfusions, or stored separately and recombined with red blood cells at some later date. Drew proved that his idea could work by setting up a blood bank at Columbia University.

SAVING SOLDIERS

As the 1930s came to a close, Charles Drew married Lenore Robbins on September 29, 1939. That same month, the nations of Europe were plunged into World War II.

TIME LINE

1904	1933	1935	1938	1941	1946	1950
Charles Drew born in Washington, D.C.	Drew graduates from McGill University.	Drew joins the faculty at Howard University.	Drew begins blood research at Columbia University.	Drew is made director of American Red Cross blood bank, but leaves in protest soon after.	Drew becomes medical director of teaching hospital at Howard University.	Drew dies.

Several months later, Drew became the first African American student to earn a doctorate in medical science from Columbia University with his work on blood banking. His expertise in blood transfusion now came into its own. His former anatomy professor, John Beattie, had returned to England, where he was working on the difficult problem of organizing blood transfusions for soldiers injured in the war. As fighting intensified, the need for blood became acute. Beattie sent Drew a telegram in New York, asking him to help the British by supplying five thousand packets of blood plasma.

Struggles for Equal Treatment

When Charles Drew died in 1950, the United States had two separate medical systems: one for blacks and one for whites. In Birmingham, Alabama, a center of many civil rights struggles, black physicians were excluded from jobs in the city hospitals until 1954. Black residents were given access to those same hospitals only a decade later, with the passage of the 1964 Civil Rights Act.

That law came into force partly through the campaigns of black doctors, who were seeking to overturn racial segregation in the medical system. One of them was Paul D. Cornely at the University of Michigan. During the 1940s and 1950s, Cornely ran a project to study the different treatment of blacks and whites in American hospitals. He carefully documented what he found and publicized the shocking inequalities in scores of articles and many public meetings. Such campaigning increased the pressure for change. In the 1950s, another Michigan graduate, Hubert Eaton, applied for a physician's post at the local hospital for whites. When he was turned down, as he expected, he filed suit and began a long series of legal challenges until he won a victory in 1964 in the federal appeals court.

Thanks to these and other determined doctors, and the earlier efforts of pioneers such as Charles Drew, that year marked the beginning of the end for racial segregation in hospitals. When the Civil Rights Act was passed in 1964, federal government funds could no longer be given to organizations or activities that segregated blacks and whites. Thus, when the Medicare program was introduced in 1966, hospitals had to end segregation or lose their federal funding. In a matter of months, one thousand hospitals quickly and efficiently ended segregation—and set out on the road to racial equality.

Within weeks, Charles Drew was organizing a large-scale lifesaving operation that became known as "Blood for Britain": blood was donated in New York hospitals, and the plasma was separated, banked, and then shipped to Britain, where it helped to save the lives of many thousands injured in the conflict. The project was an immediate and spectacular success.

As it became more apparent that the United States might enter the war, a similar operation was needed at home. Thus, in February 1941, Charles Drew was appointed director of the first American Red Cross blood bank at Presbyterian Hospital in New York City. There he was in charge of collecting, storing, and distributing blood for the U.S. Army and Navy. Drew remained in this post only nine months. A military directive required him to store blood collected from black donors separately from the blood donated by whites. Drew was appalled: skin color had no connection to blood. The order meant that injured soldiers, sailors, and airmen would have to wait for blood transfusions from those of the same racial type and that many people—black and white—could therefore die unnecessarily. Drew resigned in protest.

TEACHING DOCTORS

After leaving the Red Cross, Drew found a new way to save lives—by helping to train physicians and surgeons at Howard University. From 1941 until 1950, he was professor of medicine, and his department quickly gained a reputation for first-class training and academic excellence. He also worked as a surgeon at Freedman's, the teaching hospital linked to Howard, where he became the medical director in 1946.

During the 1940s, Drew published more than a dozen important papers on blood and transfusions, which helped to secure his reputation as an international authority. During the next few years, his expertise was recognized with a variety of honors and awards. In 1944, his work on blood plasma earned him the Spingarn Medal from the National Association for the Advancement of Colored People. The following year, he was granted honorary degrees by Virginia State College and Amherst College. He was elected a Fellow of the International College of Surgeons in 1946.

DEATH AND CONTROVERSY

On April 1, 1950, Charles Drew was driving three of his colleagues to a medical conference in Alabama. As they passed Burlington, North Carolina, he fell asleep at the wheel. The car swerved and rolled over, and Drew was thrown out of the vehicle. The car tipped on top of him, causing massive injuries to Drew's head and chest. The other doctors in the car had only minor injuries. Drew was immediately rushed to the

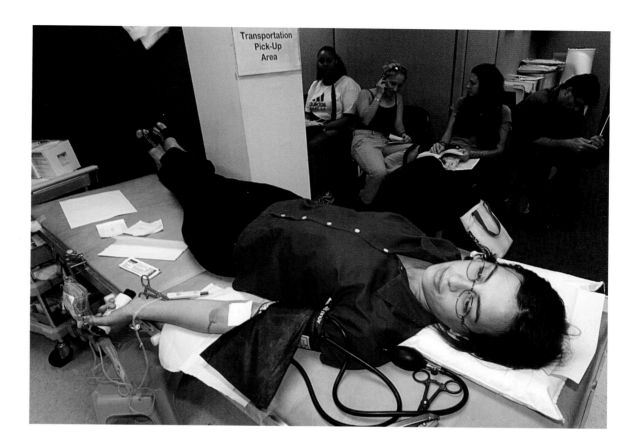

A woman donates blood to the Red Cross in Washington, D.C.

hospital in Burlington, but the doctors there were unable to save him and he died shortly afterward.

Rumors soon began to circulate that Drew had been admitted to the hospital for a blood transfusion, but was turned away because he was a black man. This would have been a supreme irony, given Drew's background and his personal battle for racial equality. It would not have been unusual, however, given the inequalities in medical care for African Americans at the time. Later, the rumor was proved to be unfounded: one of the other men traveling in the car, John Ford, confirmed that Drew was simply too badly injured to survive. According to Ford, Drew had in fact received "the very best of care" and "all the blood in the world could not have saved him."

Throughout his life, Charles Drew was committed to excellence. In his early years, he was a superb athlete. Later, he became an outstanding medical student, surgeon, doctor, and teacher. His pioneering studies of blood led to the establishment, worldwide, of blood banks from the 1940s onward. His insistence that race was irrelevant in medical treatment played a significant part in the wider struggle for civil rights during the 20th century.

—Chris Woodford

Further Reading

Book

Porter, Roy. *The Cambridge Illustrated History of Medicine*. New York: Cambridge University Press, 2001.

Web sites

Massachusetts Hall of Black Achievement
    Celebrates the achievements of black pioneers, including Charles Drew.
    http://www.bridgew.edu/HOBA/Gallery.cfm
United States National Library of Medicine
    Online exhibits about the history of medicine.
    http://www.nlm.nih.gov/onlineexhibitions.html

See also: Health and Medicine.

JAMES DYSON

Inventor of the cyclonic vacuum cleaner
1947–

Society will always need inventors. Even if there are no new problems to solve, old inventions can always be improved. Improving a classic idea can be difficult, however, as British engineer James Dyson can attest—his decision to redesign the vacuum cleaner made this clear to him. He devoted years to development before his invention was as successful as the machine it sought to replace.

EARLY YEARS

A child of academic parents, James Dyson was born in 1947 and grew up in Norfolk in eastern England. He never intended to be an inventor; in a speech in 2004, he explained his bafflement at becoming an inventor after starting out as an "artistic" youth. At school, he took no science courses—his interests focused on art and design. At age 19, he began studying furniture design and interior design at London's Royal College of Art (RCA). Then, something unexpected happened. "While at the RCA, I accidentally discovered the glories of making things," Dyson recalled later. "And I can tell you it was quite a shock when I realized I was getting interested in engineering."

In the late 1960s, Dyson designed theaters, airport lounges, and wineshops, before focusing on inventing. Working with British inventor Jeremy Fry, he developed a fast military boat, the Sea Truck, which won awards and sold in 50 countries. By 1970, he was working at Fry's company, Rotork, managing the Sea Truck project.

Shortly thereafter, Dyson set up a firm to develop his own ideas. His first big invention, in 1974, was the Ballbarrow: a wheelbarrow with a large red ball at the front instead of a wheel; this ball made it much more maneuverable over muddy ground than a traditional barrow. Later, he used the same idea to make the Trolleyball, a boat-launching trolley with balls instead of wheels. The Waterolla was another invention from this period. It was a simple garden roller (a device for flattening rough ground) that had in place of the usual metal roller a large plastic cylinder that could be filled with water, turning it into a heavy rolling drum.

REINVENTING THE VACUUM

In 2005, Dyson demonstrated the latest version of his vacuum cleaner to shoppers in London.

In 1979, at age 32, Dyson bought and started renovating an old country house and was irritated when his secondhand vacuum cleaner became clogged with dust. He discovered a similar problem at his Ballbarrow factory, where the air filter was constantly blocked with powder from the manufacturing process. Dyson designed a new machine that would clean the air by sucking in dusty air, spinning it around to remove the dust using centrifugal force (the force that pushes things outward when they spin around in a circle), and then blowing clean air back into the room.

Dyson was delighted with this machine and began to wonder why the same technique was not used in domestic vacuums. So he started tinkering with a cereal box and some masking tape, and he built a model of a centrifugal vacuum cleaner. Between 1979 and 1984, he tried no fewer than 5,127 different prototypes (test versions) of his invention until he had a product he could manufacture. During this time, he searched

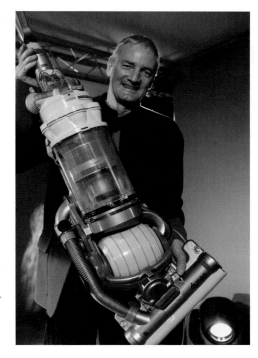

Europe in vain for investors who would finance the project, but he could not convince anyone that customers wanted vacuum cleaners to work any differently. With no luck in Europe, he took his product to Japan. Named the G-Force, it was manufactured there from 1986 and became a great success. This original model was bright pink and became a status symbol in Japan, selling for $2,000.

Dyson, however, was determined to make a product that he could sell more widely. In 1993, using money he had earned from the G-Force, he started manufacturing a new model, the Dyson DC01, in his own British factory. Even though the DC01 cost twice as much as a conventional vacuum, it had twice the suction of a normal vacuum with no messy bags to change. The DC01 became the best-selling vacuum in Britain within two years. Dyson was quickly seen as more than an inventor. In Britain, he became a national hero: he was making a successful product in his own country at a time when it seemed virtually all domestic appliances were being manufactured overseas and imported.

UPS AND DOWNS

Throughout the 1990s, sales continued to rise. Dyson's products earned design awards worldwide and he received many personal honors, including being named a Commander of the British Empire, one of Britain's highest awards. He also encountered difficulties. In 1997, he brought a case before the European Court of Human Rights to protest patent fees

TIME LINE

1947	1960s	1974	1979	1986	1993	2000
James Dyson born in Norfolk, England.	Dyson develops the Sea Truck with Jeremy Fry.	Dyson invents the Ballbarrow.	Dyson begins work on a new type of vacuum cleaner.	Dyson launches the G-Force vacuum cleaner in Japan.	Dyson begins manufacturing the Dyson DC01 in Britain.	Dyson launches a washing machine and carpet cleaner.

that inventors have to pay to protect their ideas. He also won lawsuits against the Hoover and Electrolux companies, which he claimed had copied parts of his invention and infringed his patent.

As Dyson's reputation grew, the value of the Dyson brand also grew—and he used it to sell new inventions. In 2000, he launched a washing machine, the DC06 Contrarotator, that had two different drums rotating in opposite directions. The same year, he began selling a revolutionary carpet cleaner, the DC04 Zorbster. Both products won awards.

As part of his plan to conquer the American market, he announced he would close his British factory and relocate manufacturing to Malaysia. The outraged British media claimed that 800 assembly-line jobs would be lost, but Dyson defended the decision robustly: "All our competitors were manufacturing in China, while we were watching our profits go into free fall. I could see our demise." In the end, he was able to save 210 of the 800 jobs by expanding his British research and development wing. The decision to manufacture overseas was controversial, but it was vindicated by the continuing success of Dyson's company.

—Chris Woodford

Further Reading

Books

Dyson, James. *James Dyson's History of Great Inventions*. London: Constable Robinson, 2002.
———. *Against the Odds*. New York: Thomson/Texere, 2003.
Ierley, Merritt. *Comforts of Home: The American House and the Evolution of Modern Convenience*. Mendham, NJ: Astragal, 1999.

Web site

Dyson
Includes the story of James Dyson, his inventions, and his company.
http://www.dyson.com/

See also: Cochran, Josephine; Household Inventions.

GEORGE EASTMAN

Inventor of the handheld camera
1854–1932

Although many inventors are notable scientists or technicians, George Eastman was first and foremost a businessman. He founded the Eastman Kodak Company to market his invention and became one of the wealthiest people in the United States. Historians argue that his real genius was in creating a company that fostered innovation: many of his major breakthroughs were achieved in conjunction with people he had hired. However, Eastman's inventions were also revolutionary. He transformed photography from an esoteric art form, practiced by an elite few, to an easy way for ordinary people to record and share images.

EARLY YEARS

George Eastman was born in 1854 in Waterville, New York, the youngest of three children. His father owned a business school in Rochester, New York, about 120 miles (193 km) from Waterville. In the 1860s the Eastman family moved to Rochester, where they lived well. In 1862, however, Eastman's father became ill and died, and Eastman's mother had to take in boarders to make ends meet.

In 1868 Eastman dropped out of school over his mother's objections to contribute financially to the family. Life continued to be difficult for the Eastman family, and George's polio-stricken older sister died in 1870. George Eastman worked his way up to better-paying jobs, becoming an insurance agent and then a bank clerk.

A CONSUMING HOBBY

In 1877, while Eastman was working as a bookkeeper, he heard about a real-estate boom in the Caribbean. Eastman considered visiting the Caribbean to invest in real estate, and a friend suggested that he learn

George Eastman, photographed around 1890.

photography so he could make a record of potential sites. Eastman never made the trip, but he did pay nearly fifty dollars (a large amount at a time when the average annual salary was between $300 and $400) for photographic gear, or an "outfit," as it was known at the time. The massive size and weight of the outfit troubled him; he later recalled thinking that "it seemed that one ought to be able to carry less than a packhorse load" when taking pictures.

Creating a photograph in the 19th century was a major undertaking. A photographer had to take a sheet (plate) of glass, dip it into egg whites, coat it with a variety of chemicals

Before Eastman's inventions, photographers had to carry wooden cases filled with equipment, as shown in this illustration from around 1869.

(an emulsion) to make it sensitive to light, and then load the glass into a camera, just to take a picture. Once the photograph was taken, the photographer had to develop it. The process of developing a photograph was so complicated that almost the only people who took the trouble to learn it were those planning to make a living as photographers.

Eastman, who was interested in taking outdoor photographs, faced another complication: the emulsion had to be put on the glass plates in the dark to avoid overexposure. Consequently, whenever Eastman wanted to take a photograph outside, he first had to set up a dark tent in which he could prepare his plates.

In 1878, Eastman became interested in a new tool, a dry-emulsion plate, which he had read about in a photography journal. Unlike the wet-emulsion plates he was using, a dry-emulsion plate could be prepared ahead of time and loaded into a camera at the photographer's convenience. Eastman quickly set out to make his own dry-emulsion plates. "At first," he later recalled, "I wanted to make photography simpler merely for my own convenience, but soon I thought of the possibilities of commercial production."

COMMERCIAL PRODUCTION

Eastman was not the only one who saw the commercial potential of dry-emulsion plates, which soon began to appear in stores. The plates were made by hand and were relatively expensive; Eastman thought that a machine could be used to make them less expensively. In 1879, Eastman took a leave of absence from his work and traveled to London, where he patented a machine for producing dry-emulsion plates. Eastman returned to the United States, where he patented the machine again and set up a factory to produce dry-emulsion plates. He began manufacturing Eastman-brand plates for sale while keeping his bookkeeping job.

In 1881, Eastman quit his job and began to focus on his plate business full-time. That year, he also began looking for a substitute for the heavy, breakable glass plates used to make negatives. He created a workable film using coated paper that had been treated with oil to make it transparent; in 1885, along with another photographer, Eastman developed a "roll holder" that allowed users to convert their plate cameras to ones that could use paper film.

These products did not appeal to serious photographers. Despite all the inconveniences of the traditional camera, prints made from glass plates were crisper and more detailed than ones made from paper film. Eastman realized that he was not going to convert professional photographers to his film; instead, the key to enlarging his business would be to expand the popularity of photography by making it accessible to more people.

THE KODAK

In 1888, Eastman introduced a new camera to the market, one that was remarkable in its simplicity: the Kodak. While not cheap, the Kodak was relatively inexpensive for a camera, costing $25. So small that it did not require a tripod and could be held in the hand or worn around the neck with a strap, it came preloaded with enough paper film to take one hundred photographs. Once the film was used up, the customer could mail the entire camera back to the Kodak plant, which for $10 would develop all the pictures and reload the camera with fresh film.

The name Kodak was devised by Eastman, who wanted a short, distinctive name for his camera that would be easy to pronounce and would con-

TIME LINE

1854	1878	1879	1881	1885	1888	1900	1932
George Eastman born in Waterville, New York.	Eastman becomes interested in photography.	Eastman patents a machine to produce dry-emulsion plates.	Eastman invents photographic film.	Eastman develops the roll holder.	Eastman introduces the Kodak camera.	The Brownie is released.	Eastman dies.

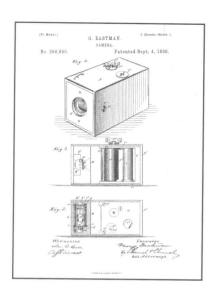

Illustrations from Eastman's 1888 patent and a photograph of one of the first cameras produced by Eastman's company.

tain the letter "k," which he liked. Eastman also devised a popular slogan touting the simplicity of his camera: "You press the button, we do the rest."

The camera caused a sensation, swiftly opening the field of photography to anyone who could push a button to take a picture. Within eight years, about one hundred thousand Kodaks were sold, a remarkable feat for a camera in the late 19th century. Kodak cameras began to show up in unexpected locales, with enthusiastic amateurs snapping pictures of people on the street or the beach—sometimes to the dismay of those being photographed. Numerous magazine and newspaper articles were written about the Kodak craze and those "kodakers" who had been swept up in it, and the name Kodak appeared in popular songs and musicals.

Some predicted that the handheld camera craze would be replaced by another fad, but Eastman was more optimistic. He observed that people were not taking photographs simply for the sake of recreation; they were taking photographs to record events in their lives, a need that would not disappear.

Under Eastman's leadership, what eventually became the Eastman Kodak Company continued to develop smaller, cheaper, easier-to-use cameras, eventually creating a $1 camera with 15-cent film—the Brownie. That camera, introduced in 1900, was marketed specifically to children, in keeping with Eastman's philosophy that cameras could be made simple enough for anyone to use.

BREAKTHROUGHS IN FILM

Although paper film was lighter and less fragile than glass plates, it was far from perfect. Paper had a grain that appeared in prints, and paper film was delicate. Eastman realized that the film market could become lucrative, as film was used once and replaced often.

As early as 1888, Eastman made it a priority to develop a tough, flexible, transparent film. The next year, one of Eastman's chemists, Henry Reichenbach, made a breakthrough, creating a workable film out of celluloid, a durable-yet-flexible material made from plant fiber. By August 1889, Eastman had the film on the market, and within months it was selling so well that the company could not produce the film fast enough.

Eastman kept on top of film technology, creating film products for X rays after their discovery in 1895. In 1889, Eastman developed film for motion pictures, then produced a fireproof safety film in 1909.

LATER LIFE

By the early 20th century, Eastman Kodak employed around four thousand people and had operations in Europe, North America, and Asia. The company would eventually come to control more than three-quarters of the total photography business in the United States. Eastman continued to run Eastman Kodak before semi-retiring in 1925. By then, he had begun to move away from laboratory work to focus on managing his increasingly massive firm.

Nonetheless, Eastman continued to recognize the importance of research to the success of his company; he recognized, too, that increased accessibility to technical and scientific education would benefit the photography industry. Eastman, one of the wealthiest people in the United

The Brownie camera, introduced in 1900, was essentially a cardboard box that could take photographs. The outside of the camera featured illustrations by children's book author Palmer Cox.

Eastman (left) with fellow inventor Thomas Edison, outside Eastman's Rochester home in 1929.

States by the early 20th century, became a major philanthropist, ultimately donating $80 million to various institutions. Among those donations were major gifts to the University of Rochester, the Tuskegee Normal and Industrial Institute (at the time one of only a few institutions of higher learning open to African Americans), and the Massachusetts Institute of Technology, which was able to build an entire campus because of Eastman's generosity.

Eastman enjoyed donating the money and planning how that money would be used. "Men who leave their money to be distributed by others [after they die] are pie-faced mutts," he once declared. "I want to see the action during my lifetime."

During the late 1920s and early 1930s, however, Eastman's health declined. By 1932, he was unable to go to the office or travel, he was often in pain, and his staff tried to protect him by discouraging visitors. Eastman reportedly became isolated and depressed, feeling that he had nothing left to contribute and nothing to look forward to. On March 14, 1932, at the age of 77, Eastman committed suicide by shooting himself in the heart.

A PHOTOGRAPHIC WORLD

The Eastman Kodak Company survived its founder's death; in 2005 the company sold $14 billion worth of photographic equipment and supplies. Although it is still a large corporation, Eastman Kodak no longer dominates the industry as it once did. With the advent of digital photography,

Eastman's dream of making photography accessible to ordinary people has been fully realized in the 21st century. Digital photography has made it easy for people to take and print their own photographs.

computer, software, and even telephone companies have entered the field of photography.

Eastman did more than help to create the photography industry, however; he popularized photography. Photography has become so much a part of modern life that it would be unusual to not take pictures of family members or notable events. Photography has also become a serious art form, and photographic images have defined events and profoundly influenced people's perceptions of the world.

—Mary Sisson

Further Reading

Books

Ackerman, Carl W. *George Eastman.* Boston: Houghton Mifflin, 1930.

Brayer, Elizabeth. *George Eastman: A Biography.* Baltimore, MD: Johns Hopkins University Press, 1996.

Web sites

George Eastman House
> Exhibits from the museum that currently occupies Eastman's home.
> http://www.eastmanhouse.org/

History of Kodak
> Information on Eastman and his company by the Eastman Kodak Company.
> http://www.kodak.com/US/en/corp/kodakHistory/index.shtml

The Wizard of Photography
> An exhibit on Eastman from PBS's *The American Experience.*
> http://www.pbs.org/wgbh/amex/eastman/

See also: Baekeland, Leo; Daguerre, Louis; Edison, Thomas; Entertainment; Lumière, Auguste, and Louis Lumière.

THOMAS EDISON

Prolific inventor

1847–1931

A glowing lightbulb is often used to symbolize a great idea. The electric lightbulb was just one of 1,093 great ideas patented by Thomas Edison, whom many regard as the finest inventor of all time. What sets Edison apart is not just the great number of his ideas but also their significance. At least three of his inventions—practical electric power, sound recording, and moviemaking—created fundamental social changes during his lifetime. Because Edison pioneered the concept of the modern research laboratory, his work has continued to have a major influence on inventors long after his death.

EARLY YEARS

Edison was born on February 11, 1847, in the Great Lakes port town of Milan, Ohio; he was the last of seven children of Samuel Ogden Edison and Nancy Matthews Elliott, his wife. When Edison was seven, the family moved to Port Huron, Michigan, where his father became a grain merchant. Edison began school in Port Huron—an education lasting just 12 weeks.

From an early age, Edison had difficulty hearing. At the time, most learning was by rote (children had to listen to their teachers and repeat exactly what they said). Edison's hearing problem made this impossible in a class full of noisy young children; his teacher, a minister named Engle, branded him a misfit. Observing that Edison's head was large and strangely shaped, Engle suggested that the boy's brains were "addled" and defective.

> Opportunity is missed by most people because it is dressed in overalls and looks like work.
>
> —Thomas Edison

Edison's mother, a former schoolteacher, decided to teach him at home. Edison later wrote: "My mother was the making of me. She was so true, so sure of me; and I felt I had something to live for, someone I must not disappoint." His mother gave him a love of learning, and he eagerly read everything from science books to Shakespeare's plays, from poetry to world history.

Edison's father was reasonably wealthy, and as a young boy Edison did not need to work. However, he wanted independence. In 1859, he took a job selling newspapers and candy on trains that ran on the Grand Trunk Railroad. Three years later, at age 15, he started compiling his own weekly newspaper, printing it on a small handpress in the baggage car, and selling it on the train. This gave Edison a taste of fame: England's newspaper, the *Times*, reported the venture as the first newspaper ever to be printed on a moving train.

BECOMING AN INVENTOR

One day, while selling papers on the train, Edison saved a boy's life. The child's grateful father, who worked at the train station, rewarded Edison by teaching him how to operate a telegraph. This invention used electric cables to carry messages back and forth at high speed, one letter at a time, written in Morse code, a special pattern of short (dots) and long (dashes) electrical pulses. Learning about the telegraph changed Edison's life. In 1862, he started working as a telegraph operator in a bookstore in Port Huron; the following year, he operated telegraphs along the Grand Trunk Railroad in Ontario, Canada, before returning to the United States.

Thomas Alva Edison, photographed around 1872.

Over the next five years, Edison worked his way from one telegraph office to another. This experience gave him an idea for his first invention, the automatic telegraph repeater. It could carry messages in Morse code through unmanned offices to stations farther down the line. Despite the usefulness of the invention, Edison was reprimanded by his boss, who believed the young inventor was wasting company time.

Edison moved to Boston in 1868 and to New York City the following year. In New York, at age 21, he invented an electric machine that could count votes in elections. Unfortunately, that invention never took off, but Edison's flair for inventing soon earned him rewards. While working for a financial company, he was asked to improve its "stock ticker" (a printer that recorded stock prices arriving as telegraph messages) and was paid a large sum.

Edison made many improvements to the telegraph in the 1870s, and he set up several companies to implement them. One of his greatest achievements was devising a method to send four different messages along a single telegraph wire concurrently. He called this invention the "quadruplex telegraph" and sold it to Jay Gould, American businessman and so-called robber baron, for $30,000 (about $500,000 today). This achievement marked Edison's arrival as a successful inventor.

On Christmas Day 1871, Edison married Mary Stilwell, a woman who had worked for one his companies. They had three children; the first two, Marion and Thomas Jr., he affectionately nicknamed "Dot" and "Dash."

THE WIZARD OF MENLO PARK

In 1876, Edison used his money from the telegraph to set up a research laboratory at Menlo Park, New Jersey, where he developed some of his greatest inventions. Edison's laboratory was a messy, creative place with

Edison's Menlo Park laboratory as reconstructed in the Henry Ford Museum in Dearborn, Michigan.

no rules: dirt was allowed to gather on the floor, workers could spit freely, and a chained-up pet bear stood guard outside—until it broke free in April 1877 and had to be killed.

Later that year, Western Union persuaded Edison to develop a new telephone that could compete with the one that Alexander Graham Bell had invented. Soon, Edison developed a carbon microphone that captured sound more clearly than Bell's. During this research, Edison found he needed a way to signal the beginning of a telephone conversation to someone at the other end of the line and hit upon "Hello"—another of his inventions that became universal.

Edison's telephone work led to another breakthrough later the same year. He had already developed machines to help telegraph operators record incoming messages; now he wondered if he could produce something similar for the telephone. This led him to invent the phonograph, the world's first sound-recording machine (see box, The Phonograph). The phonograph secured Edison's fame. A reporter from a newspaper, the *New York Graphic*, came to interview Edison at Menlo Park and commented: "Aren't you a good deal of a wizard, Mr. Edison?" When the paper ran the interview, it carried a picture of Edison wearing a wizard's hat on its front cover, and his nickname, "the wizard of Menlo Park," was born. The same paper ran an April Fools' joke describing a machine Edison had supposedly invented that could automatically turn soil into cereal and water into wine.

If the phonograph proved that Edison could be original, his next invention demonstrated his more practical side. During 1878 and 1879, after filling 40,000 pages of notes with scribbled ideas, Edison developed an improved electric lightbulb. It consisted of a thin piece of material (filament) that glowed red hot and gave off light when electricity passed through it. Earlier light designs had worked in a similar manner, but their carbon filaments quickly burned up in the air, so they were never practical. To find the perfect filament, Edison experimented with more than 6,000 different materials, including

An advertisement for an Edison Triumph phonograph from around 1901.

The Phonograph

"Mary had a little lamb,
its fleece was white as snow. . . ."

Edison's employees could not believe what they were hearing. Some thought it was a magic trick; others insisted there was a ventriloquist hiding in the room, making the sounds. Either way, it seemed unbelievable: as they huddled around the little machine, one day in 1877, they distinctly heard Edison's voice crackling out of the horn: "Mary had a little lamb, its fleece was white as snow." This "little piece of practical poetry," as Edison called it, was the world's first sound recording. The machine that stored the sound and played it back was Edison's phonograph, the ancestor of every modern sound recording technology, from the vinyl record and the magnetic cassette to the sampling keyboard and the iPod.

Edison did not originally see the phonograph as a way of storing and playing back music. He imagined people would use phonographs in their homes, as telephone answering machines or in offices, as aids to dictation. Another early use he imagined was in toys. He formed a company to manufacture talking dolls with windup speaking cylinders inside them, which could automatically recite nursery rhymes. The phonograph was not an immediate success, however, and Edison lost interest in the idea for almost a decade while he devoted himself to electric light and power.

Edison listens to his phonograph through primitive headphones.

Returning to the idea in 1887, he developed an improved phonograph using wax cylinders instead of foil sheets. During the 1890s, Edison released a whole series of improved phonographs, with a standard model costing around $20 (about $450 today). In the early 1900s, his companies began selling mass-produced wax cylinders, each capable of storing two to four minutes of prerecorded sound or music, and costing from 50 cents to $4.

Edison's rivals initially copied his idea, but they later switched to flat circular discs that could be played on a similar machine called a "gramophone." When in 1912 Edison started selling his own version, the disc phonograph, the days of the cylinder phonograph were numbered.

How Edison's Phonograph Worked

1 Operator cranks the handle.
2 Heavy metal cylinder rotates.
3 Tinfoil attached to cylinder surface rotates with the cylinder.

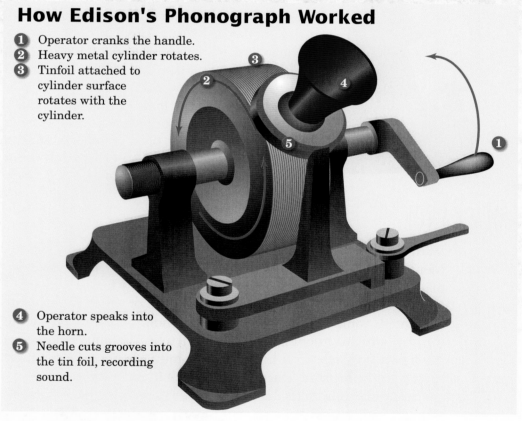

4 Operator speaks into the horn.
5 Needle cuts grooves into the tin foil, recording sound.

Edison's original phonograph was entirely mechanical—it used no electricity. At its heart was a small metal cylinder that rotated when the operator cranked a wooden handle. The cylinder had a thin sheet of foil wrapped around it, on top of which a horn with a sharp needle on the end rested. To record sound, Edison spoke into the horn and cranked the handle so the cylinder rotated. The sound of his voice made the needle vibrate up and down, cutting a spiral groove into the foil. To play back the sound, Edison moved the needle and horn away from the foil and put it back at the start of the spiral groove. When he turned the crank, the needle followed its original path, bumping up and down in the groove, and "reading back" the recorded sound, which the horn amplified (made loud enough to hear).

An electric light, made by Edison in 1879.

cotton, bamboo gathered specifically for that purpose in Japan, and even red hair, before settling on wire made from tungsten metal. Edison also placed his filament inside a glass vacuum bulb, extending its life by hundreds of times. Although Edison did not invent the electric light, he turned it into the practical invention that lit up people's homes.

From 1878 onward, Edison moved to capitalize on electricity: his idea was to sell electric light to the nation and later to the world. However, before he could do so, he had to provide people with electric power. Thus, he began designing dynamos (coal-powered electricity generators). He nicknamed his first one "Long-Legged Mary Ann," in honor of its two tall magnets, which stood upright like a gigantic pair of legs. Machines like these formed the heart of the world's earliest power plants, which Edison built during the 1880s. He opened the first of these on Pearl Street, in New York City, on September 4, 1882, supplying electricity to 59 customers in lower Manhattan.

TIME OF CHANGE

With his work now centered on New York City, Edison closed his Menlo Park laboratory. Working from offices at 65 Fifth Avenue, he maintained a feverish pace. In 1882, he applied for 141 patents (an average of two to three a week), just over half of which were eventually granted. Some days he worked 18 or 19 hours, unaware of whether it was day or night, stopping to eat only when he was hungry—and spending hardly any time with his wife and family. He once said, "I owe my success to the fact that I never had a clock in my workroom."

Still only in his mid-thirties, Edison was securing his reputation as a great inventor and the father of the electrical age. In 1884, Edison's wife, Mary, died of typhoid fever, leaving him to bring up three small children. Two years later, he married Mina Miller; they had three children, including Charles—who as an adult would run his father's businesses and be elected governor of New Jersey. Edison acquired a large estate, Glenmont, in West Orange, New Jersey, for his expanding family and, in 1887, built himself a huge new research laboratory—or "invention factory," as he called it—nearby.

The Battle for Electric Power

Edison's brilliance made him many friends, such as Henry Ford, Marie Curie, Herbert Hoover, and Charles Lindbergh. It made him enemies, too, none more so than Croatian-born inventor Nikola Tesla (1856–1943). Tesla settled in 1884 in New York City, where he was employed by Edison, who was busy bringing electricity to the nation. However, the two men could not get along, so Tesla quit the following year and struck out alone.

Edison and Tesla were prolific inventors (Tesla eventually gained 100 patents from a total of around 700 different inventions) and visionaries who could see that electricity would power the future. Yet they were also bitter rivals who disagreed on a fundamental issue. Edison believed in a type of power called "direct current" (DC), in which electricity moves continually in the same direction.

Tesla thought alternating current (AC) was better. In this system, the electric current reverses direction (alternates) many times each second. The advantage of alternating current is that it can be transformed (boosted electrically) to a much higher voltage and then be carried many miles from a power station to where it is needed with much less loss of power than with direct current.

After leaving Edison's firm, Tesla worked by himself to develop his AC power system, completing its design around 1887. The following year, he sold the rights to the Westinghouse Electrical Company. Edison did his best to fight the AC system with canny publicity stunts. For example, he actively campaigned for the death penalty to be administered by an AC-powered electric chair, because he thought this would convince people that AC was a lethal technology. As part of his battle, he staged outrageous demonstrations for the press: he killed cats, dogs, a horse, and even an elephant called Topsy to demonstrate the dangers of AC electricity—leading the press to coin a new word, electrocution.

Edison won the publicity battle, but Westinghouse won the war. It was awarded the contract to build a giant power plant at Niagara Falls using Tesla's AC system in the 1890s, defeating a rival bid from Edison's DC system. Since then, virtually all of the world's electric power has been supplied using AC.

West Orange marked a turning point in Edison's life. At Menlo Park, he had worked with a small team on a few ideas. However, at West Orange, more inventions were in development and more people were involved. Edison was also increasingly moving from the role of inventor to industrialist: he was no longer merely thinking of ideas, but now running companies to profit from them. The West Orange lab was a more businesslike place than Menlo Park. It was run by managers and administrators, workers had to punch a time clock, and Edison became more aloof, directing the lab rather than working in it himself.

During this period, Edison began to experience some serious commercial failures. In 1887, he developed a way of using electricity and magnetism to separate gold and iron from their ores (rocks dug from the ground). He spent more than a decade and a large amount of money on this project, bought 145 old mines, and developed a gigantic and hugely expensive plant in New Jersey. Although he expected to make a fortune, he never perfected the process and was forced to close the plant in the 1890s at a loss of $2 million. Edison experienced another major disappointment at around the same time when the system he used for supplying electric power (known as direct current) was shown to be inferior to a rival system, alternating current, promoted by the Westinghouse Electrical Company (see box, The Battle for Electric Power).

MAKING MOVIES

The kinetoscope, invented by Edison and William Dickson. To see the movie, a viewer had to look into the eyepiece at the top of the machine.

While trying to think of ways to promote the phonograph, Edison hit upon his next big idea: the movie camera and projector. In the 1880s, he had seen demonstrations of toys that could make images appear to move. They worked by spinning a series of still pictures on a wheel until the images blurred together in the viewer's eye to make a moving scene. With his assistant William Dickson (1860–1935), Edison developed an invention that took this idea a step further. It was a primitive movie

An 1896 advertisement for an early film projector called the vitascope.

camera, called the kinetograph, that could record a series of still photographs on to a length of celluloid film. The same filmstrip could later be run at high speed through a piece of apparatus called a kinetoscope so people could view the "movie."

In 1893, Edison opened the world's first studio, nicknamed the "Black Maria," and began filming short motion pictures. Because the concept of the movie theater had not yet been brought to realization, Edison's movies could be viewed by only one person at a time. Viewers had to peer inside coin-operated kinetoscopes in large parlors where dozens of machines were installed side by side. Soon after, two French brothers, Auguste and Louis Lumière (1862–1954; 1864–1948), saw some of Edison's machines. When they figured out how to make a better camera that could project images onto a wall, modern movies were born.

LATER YEARS

Much of Edison's time was devoted to developing three ideas: electric light and power, sound recording (the phonograph), and movies (the kinetograph, kinetoscope, and later a projector called the vitascope). However, he still found time to explore other ideas. In the 1880s and

TIME LINE

1847	1862	1870s	1876	1877
Thomas Edison born in Milan, Ohio.	Edison begins work as a telegraph operator.	Edison makes several improvements to the telegraph.	Edison sets up a research laboratory in Menlo Park, New Jersey.	Edison invents the carbon microphone and phonograph.

1890s, he developed large electric batteries. He originally designed these to help his friend Henry Ford (1863–1947), the automobile pioneer, who wanted to use electricity in his cars. Edison and Ford dreamed of using batteries to make an electric-powered car, but their creations never succeeded in rivaling the gasoline engine. In 1910, Edison demonstrated a way of making a home from poured concrete, receiving great acclaim.

During World War I, when a journalist asked Edison how the United States could win wars, he suggested that the U.S. government "should maintain a great research laboratory" like his own. In 1915, the government asked Edison to head the Naval Consulting Board, a body that would advise the navy on using the latest technology. During this period, Edison made many discoveries helpful to the military, including a new way of launching torpedoes and a method of manufacturing

Around 1929, Edison uses his invention, the Telescribe, a kind of dictating machine, to give instructions to an employee.

TIME LINE

1879	1882	1893	1926	1931
Edison invents the electric lightbulb.	Edison opens the first electric power plant in New York City.	Edison opens the first movie studio.	Edison resigns as president of his company.	Edison files his last patent application and dies later that year.

synthetic rubber (to avoid being dependent on imports of natural rubber during the war). In 1923, Edison's work led Congress to set up the Naval Research Laboratory, which has remained one of the world's most important centers of military innovation.

Edison remained active well into his seventies, and his pace of work slowed only when his health began to fail. When he resigned as president of his company in 1926, his son Charles took over. Edison filed his last patent application on January 6, 1931, and died in West Orange nine months later on October 18. Just before his death, he woke briefly from a coma and uttered his last words to his wife: "It is very beautiful over there." A few days later, every electric light in the country—including the one that powered the torch in the Statue of Liberty—was turned off for one minute in tribute.

EDISON'S IMPORTANCE

Edison's inventions had a huge impact on people's lives. More than any other individual, he helped to usher in the modern age of electric power, convenience, and entertainment. His filament light demonstrated the usefulness of electricity, opened the door to electric power, and encouraged others to develop more electrical appliances. While Edison's electricity changed homes and businesses, his work on sound recording and the movies improved people's social lives, and his telegraph and telephone improvements brought advances in communication.

Although Edison's inventions were often technical, he was more of a hands-on experimenter than a scientist or theoretician. He made no secret of this: "I try an experiment and reason out the result, somehow, by methods which I could not explain." Driven by the old saying that "necessity is the mother of invention," he always ensured that his inventions met people's needs, and he continued to perfect his inventions long after they had been launched. All this required immense perseverance, a trait he acquired during the early years when his mother had encouraged him to overcome his hearing disability with hard work. Indeed, his most famous saying was, "Genius is one percent inspiration and 99 percent perspiration."

Edison pioneered invention laboratories where scientists and engineers could work together in a commercial environment, developing ideas with immediate market potential. His "invention factories" at Menlo Park and West Orange were devoted to applied research, rather than to producing interesting ideas for which someone might find a use later. This style of inventing is the one that most companies still use, and it is one of Edison's greatest legacies: he was not just a great inventor, but a pioneer who helped to change the way people invent.

—Chris Woodford

Further Reading

Books
Baldwin, Neil. *Edison: Inventing the Century.* Chicago: University of Chicago Press, 2001.
Carlson, Laurie. *Thomas Edison for Kids: His Life and Ideas, 21 Activities.* Chicago: Chicago Review Press, 2006.
Gutman, Dan. *Back in Time with Thomas Edison.* New York: Aladdin, 2002.
Israel, Paul. *Edison: A Life of Invention.* New York: Wiley, 2000.
Sullivan, George. *Thomas Edison (In Their Own Words).* New York: Scholastic, 2002.
Time editors. *Thomas Edison: A Brilliant Inventor.* New York: HarperTrophy, 2005.

Web sites
Edison Birthplace Museum
   A museum in Milan, Ohio, dedicated to Edison's life.
   http://www.tomedison.org/
Inventing Entertainment
   A collection of Edison's sound recordings and movies at the Library of Congress.
   http://memory.loc.gov/ammem/edhtml/edhome.html
Thomas A. Edison Papers
   Edison's collected papers at Rutgers University.
   http://edison.rutgers.edu/

See also: Bell, Alexander Graham; Communications; Corporate Invention; Energy and Power; Entertainment; Ford, Henry; History of Invention; Lumière, Auguste, and Louis Lumière; Morse, Samuel; Tesla, Nikola.

ROBERT EDWARDS AND PATRICK STEPTOE

Inventors of human in vitro fertilization
1925– and 1913–1988

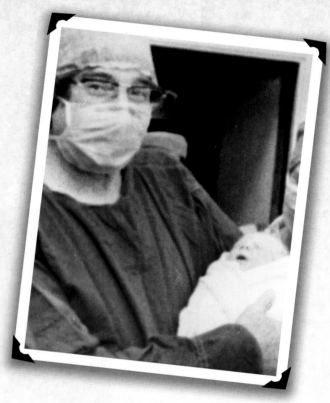

Robert Edwards and Patrick Steptoe made headlines in 1978 when Steptoe delivered Louise Brown, the first so-called test-tube baby. Brown had been conceived in a laboratory using a process called in vitro fertilization (IVF). Although enormously controversial at the time (and still condemned by some religious groups), IVF gave rise to a variety of other reproductive technologies, offering hope to infertile couples that they would be able to bear children.

EARLY YEARS: ROBERT EDWARDS

Human IVF was the idea of Robert Edwards. Born in 1925 in Leeds, a town in northern England, Edwards grew up in the industrial city of Manchester. His family had very little money, but Edwards was a bright student, and when he was 11, he was offered a scholarship to a prestigious school that his mother insisted he accept.

Following the outbreak of World War II in 1939, Edwards, along with many other British children and teenagers, was evacuated to the countryside because of German bomb attacks on British cities. Edwards would later credit the experience with piquing his interest in the natural world. When he was old enough, Edwards joined the army, serving until 1949.

After his discharge, Edwards attended college at the University of Wales at Bangor. Initially majoring in agriculture, he switched to zoology in his senior year. Upon hearing of a friend's plans to do postgraduate study in genetics at Edinburgh University, Edwards decided to follow suit. He applied and was accepted, earning a doctorate in 1955.

ATTEMPTING IVF

While at Edinburgh, Edwards studied reproduction in mice; he continued working in the field of animal reproduction for the next five years.

By 1960, Edwards had started a family and was living in London, where he worked as a researcher. His friends included an infertile couple who liked to come over and visit Edwards's daughters.

Prompted by the couple's plight, Edwards began to consider the possibility of using animal reproduction techniques to help infertile people. At the time, researchers were able take eggs from a female animal, fertilize them in a lab, and then transplant the resulting embryo into the animal's womb. When successful, the procedure, known as in vitro fertiliza-

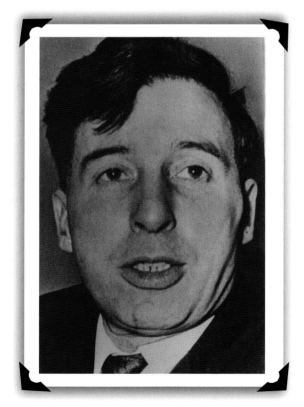

Robert Edwards, photographed in 1969.

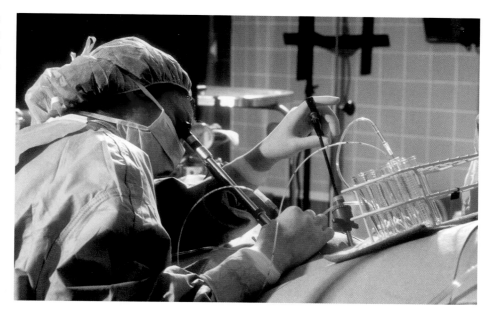

Doctors perform a laparoscopy on a female patient in preparation for an IVF procedure.

tion, would result in pregnancy and birth. Edwards thought that he might be able to perform the same procedure with humans.

Edwards, who had moved to Cambridge University in 1963, began working on fertilizing a human egg in the lab. At that time, the general belief was that human fertilization could happen only inside the body; in 1965, however, Edwards succeeded in fertilizing a human egg outside the body. When he tried to repeat the experiment, however, he failed. By 1967, Edwards had still not fertilized another egg in the lab. He decided that human sperm must somehow be altered during sexual intercourse to make fertilization possible. The only way to obtain and study such sperm would be to ask a female volunteer to submit to dangerous abdominal surgery.

In the fall of 1967, however, Edwards happened upon a journal article describing a new surgical technique—laparoscopy, which uses a thin fiber-optic tube (a laparoscope) inserted into the abdomen to provide light for special tube-shaped surgical instruments. Only small incisions were made during the procedure, making laparoscopy far less traumatic to patients than traditional surgery. The author of the article on laparoscopy, and a pioneer in the use of this surgical technique in Great Britain, was Patrick Steptoe, a gynecologist based in Oldham.

EARLY YEARS: PATRICK STEPTOE

Patrick Steptoe was born in 1913 in Whitney, England. His father was a church organist, and Steptoe himself was an adept piano player. His mother was a proponent of women's rights, and she instilled in Steptoe

Patrick Steptoe, photographed in 1970.

a profound interest in the welfare of women.

After graduating from secondary school, Steptoe attended St. George's Medical School in London, finishing in 1939, just at the beginning of World War II broke out. He volunteered for the navy. During the war, his ship was torpedoed off the island of Crete. Steptoe swam for two hours until he was captured by the Italian navy; he spent two years as a prisoner of war. After the war, Steptoe returned to London, where he worked until 1951. Deciding that it was too difficult to become professionally established there, he moved to Oldham, a fairly remote town in northern England.

Oldham had few medical professionals, and for years Steptoe struggled to provide even basic care to women in the area. Steptoe began seeing more and more cases of infertility. At the time, exploratory surgery was commonly done to determine why a woman was infertile. Steptoe sought a less extreme alternative to solving issues related to infertility. In 1959, he heard of the new laparoscopic technique, which had been developed in France. He obtained a laparoscope and eventually became one of Great Britain's foremost experts on the device, publishing a textbook on the subject in 1967.

JOINING FORCES

That same year, Edwards called Steptoe and explained how he would like to use laparoscopy to perform IVF in people. Unlike most of the gynecologists Edwards had spoken to, Steptoe was happy to help. Edwards, however, realized that working with Steptoe would be extremely challenging—Oldham and Cambridge were more than 160 miles (257 km) apart, and only small, winding country roads connected them. Edwards did not call Steptoe again.

The next year, Edwards attended a medical conference, where he ran into Steptoe. Steptoe was still excited by Edwards's idea, and Edwards realized that if he wanted human IVF to move forward, he needed someone with Steptoe's expertise. He made the first of what would prove to be countless round trips to Oldham later that year, eventually setting up a lab in the hospital where Steptoe worked.

CONTROVERSY

In Cambridge, a student of Edwards's developed a culture fluid that promoted fertilization in hamster eggs. The fluid also worked with human eggs. Although this culture fluid settled the problem of fertilization, Steptoe continued to play a crucial role in the future of IVF: his infertility patients volunteered to provide Edwards with eggs, which Steptoe

Implantation

For much of the 1970s, Edwards and Steptoe were stymied by implantation, which occurs easily in nature. When an egg is fertilized, the resulting embryo implants itself in the uterus, which has built up a thick, blood-rich lining to support it. When Edwards and Steptoe tried to implant embryos in patients, however, the implant would fail and a pregnancy would not occur.

The cause of the pregnancy failures was eventually attributed to the fertility drugs Edwards and Steptoe would give the patients. Such drugs would cause several eggs to mature at once, making retrieval with a laparoscope easier. The drugs, however, would also accelerate the mother's menstrual cycle; by the time the embryo was returned to the womb, the uterus was about to shed the lining that the embryo needed to sustain itself.

At first, Edwards and Steptoe simply modified the drug regime. In 1975, one of their patients conceived, but success proved fleeting. Steptoe determined that the embryo had implanted itself in a fallopian tube, not the uterus. Such a pregnancy would have been lethal for both the mother and the baby, and the pregnancy had to be terminated.

After two more years of failure, Edwards decided that they should not use fertility drugs at all, especially since Steptoe was skilled enough with the laparoscope to retrieve single eggs. Instead, patients were constantly monitored for the hormonal changes indicating natural ovulation, and Steptoe's team was placed on call around the clock to remove the eggs whenever they were produced.

retrieved through laparoscopy. Edwards eventually devised a culture in which the fertilized eggs could develop into embryos. The next step was to actually implant the embryo into a woman's womb to create a pregnancy, but this procedure would require a better-equipped lab than the one in Oldham.

Edwards and Steptoe applied for funding from the British government to outfit a lab near Cambridge. Edwards had been publishing papers about his progress, however, and human IVF was proving to be an extremely controversial idea. It was decried by religious leaders as an attempt by men like Edwards to take on the role of God. In 1969 a survey of Americans found that more than two-thirds thought that such fertilization technology would mean "the end of babies born through love." Other scientists even charged that IVF would result in horribly deformed children.

In 1971, Edwards and Steptoe received their reply from the British government: their attempts to develop human IVF were considered ethically suspect and unsafe, and they would not receive funding. Undaunted, the two eventually received private money and established Edwards's lab in another hospital near Oldham.

LOUISE BROWN

Lesley Brown lived in Bristol with her husband, John. After years of trying to conceive, she was discovered to have blocked fallopian tubes, organs that carry the egg from the ovary to the uterus. Surgery to fix the

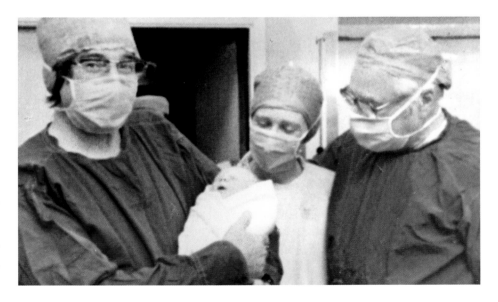

Edwards (left) holds Louise Brown shortly after her birth on July 25, 1978; Lesley Brown is pictured center and Steptoe is on the right.

TIME LINE

1913	1925	1959	1963	1968	1977	1978	1988
Patrick Steptoe born in Whitney, England.	Robert Edwards born in Leeds, England.	Steptoe begins mastering laparoscopic technique.	Edwards begins work on fertilizing a human egg.	Steptoe and Edwards begin working together.	Lesley Brown becomes pregnant through IVF.	World's first IVF baby, Louise Joy Brown, is born.	Steptoe dies.

tubes did not correct her condition. She became depressed, and her doctor finally wrote to Steptoe seeking help. In late 1977, Brown became pregnant through IVF.

Lesley Brown's pregnancy was tremendously exciting to Steptoe, Edwards, and the Browns. It was equally exciting to the British media, which soon found out that the controversial experimental procedure might actually be working. The Browns found themselves hounded by reporters, and Lesley Brown had to be admitted to the Oldham hospital under a false name.

The stress was especially troubling because late in her pregnancy, Brown developed toxemia, a potentially life-threatening condition involving bacteria in the blood. Steptoe was faced with a precarious situation: toxemia could be deadly to the mother, but treatment by inducing birth could be harmful to the baby if it had not developed fully.

Finally on July 25, 1978, Steptoe decided that the time was right for Brown's baby to be born. To avoid reporters, he arranged for a cesarean section to take place in the middle of the night. At 11:47 p.m., Louise Joy Brown, the world's first baby conceived by IVF, was born.

FROM THE EXTRAORDINARY TO THE EVERYDAY

Louise Brown was in excellent health (as of 2006, she continues to be well, as is her younger sister, who was also conceived in vitro). Six months later, a second test-tube baby was born, showing that the first success had not been a fluke. Edwards and Steptoe continued to refine their techniques, and in 1987, a year before Steptoe died of cancer, they announced the birth of their thousandth baby fertilized in vitro. By 2006, hundreds of thousands of children were believed to have been conceived and born through IVF worldwide.

Despite its success, IVF remains somewhat controversial. Certain religious groups still condemn the process, considering it unnatural or decrying the discarding of excess embryos. Also, IVF has practical

Louise Brown, center top, poses with other children born through IVF at a party held in 1993 at Columbia-Presbyterian Hospital in New York City.

shortcomings: an IVF procedure can cost upwards of $10,000—a cost that Edwards has publicly condemned. To date, the procedure has only about one chance in four of succeeding. It has also been linked to complications like birth defects and risky multiple births.

Nonetheless, IVF and the newer technologies that have developed from it have provided countless couples the opportunity to bear a child. Edwards and Steptoe's work also was a conceptual breakthrough, greatly increasing scientists' understanding of how conception and pregnancy work in human beings.

—Mary Sisson

Further Reading

Books

Edwards, Robert, and Patrick Steptoe. *A Matter of Life: The Story of a Medical Breakthrough.* New York: Morrow, 1980.

Henig, Robin Marantz. *Pandora's Baby: How the First Test Tube Babies Sparked the Reproductive Revolution.* Boston: Houghton Mifflin, 2004.

Web site

Test Tube Babies
History and information on IVF from PBS's *The American Experience* series. www.pbs.org/wgbh/amex/babies/

See also: Campbell, Keith, and Ian Wilmut; Health and Medicine.

ENERGY AND POWER

Civilization is a long and ingenious struggle by humans to survive the challenges of the natural world. Almost everything that people do requires energy of some sort; the story of humanity's struggle with nature is also the story of how people have harnessed energy in increasingly sophisticated ways. From the simple stone tools made and used by prehistoric tribes to the nuclear reactors and solar cells employed in modern times, energy inventions have given humans the power to produce goods and food, run offices and factories, and make homes warm and comfortable.

HUMAN AND ANIMAL POWER

In the 21st century, most of the energy we use comes from electricity, generated in power plants or stored in batteries, or from fuels such as gasoline and diesel pumped into trucks and cars. In the prehistoric age, however, such forms of energy were unknown, and the most sophisticated power plant was still the human body. Archaeologists believe people first used tools millions of years ago during the Stone Age, which lasted from around 600,000 BCE to around 3000 BCE. Simple tools, such as axes, picks, and blades, reduced the effort that humans needed to accomplish tasks. These tools were among the earliest inventions and represent humanity's first real attempt to use the human body's power more efficiently.

By far the most important ancient invention for harnessing energy was the wheel, developed around 3500 BCE in ancient Mesopotamia (present-day Iraq and Syria). Wheels were first used on carts to make human and animal energy

An undated woodcut of a horizontal treadmill, which provided power for a copper mine in the sixteenth century.

more effective. The bigger the wheel, the more it acted like a lever, magnifying the force that pulled a cart, reducing both the effort and the time needed to move a heavy load. A few thousand years after wheels were invented, the ancient Greeks improved them by developing gears: a pair of wheels with teeth around their edges. When the teeth interlocked, one wheel turned the other with more speed or force, depending on which wheel had more teeth. Since gears increased speed and force, they permitted one person to do as much work as several and to do the work more quickly. Although gears were invented by the Greeks, they were not widely used until Roman times (27 BCE–395 CE). The Romans made much use of animals to drive mills and presses: oxen and horses walked in circles, pushing or pulling

shafts that then turned mill wheels. The Romans also pioneered treadmills, which looked like giant hamster exercise wheels, inside which slaves or animals trudged for hours on end.

WINDMILLS AND WATERWHEELS

Generating power with humans and animals had limitations: both people and livestock required food, and they could not work without taking time to rest. In the Middle East, around 600 CE, people found a way of overcoming this problem by harnessing the power of the wind. Windmills became popular in Europe during the Middle Ages, which lasted until around 1500 CE. They were largely replaced, however, by a more dependable source of power, the waterwheel, in places with rivers and streams nearby.

Waterwheels, large open-paddle wheels made from wood that dipped into rivers, were turned by the flowing water, thereby harnessing the water's power for use in a mill or other machine. Originally invented by the Greeks, the first waterwheels were mounted horizontally. During Roman times, a distinguished architect, Vitruvius (ca. 70–25 BCE), realized that waterwheels would work better standing upright with water flowing beneath them. These were known as undershot waterwheels. At the end of the Roman period, an even better design was developed. Known as the overshot waterwheel, it was powered by water flowing down through a tray over the top of the wheel. This type of wheel could work even with slow-flowing water and

In Hit, Iraq, boys fish along the banks of the Euphrates River near an ancient waterwheel.

extracted much more energy. From Roman times until the 18th century, waterwheels were the engines of civilization.

STEAM ENGINES

Although windmills and waterwheels were great advances in harnessing energy, their power was still limited: windmills worked best in stormy weather, and waterwheels had to be situated near rivers and streams. A way of powering machines reliably, independent of circumstances, was needed. That invention finally arrived early in the 18th century, when Englishman Thomas Newcomen (1663–1729) developed the steam engine: a device that burned coal, a plentiful material dug from the ground, to drive machines. Initially used for draining wastewater from coal mines, steam engines were able to do the same work as dozens of horses. Their chief advantage was that they could work day and night and, though they consumed enormous amounts of coal, they needed neither wages nor food.

From the 18th century to the late 19th century, steam engines were the most important source of power for transportation. After English inventor George Stephenson (1781–1848) developed the first practical steam locomotive in 1814, steam engines powered a huge, worldwide expansion in railroads. Engineers such as the American Robert Fulton (1765–1815) and the Englishman Isambard Kingdom Brunel (1806–1859) pioneered the use of steam engines on large ships.

Inventors who tried to use steam engines to power other forms of transportation were largely unsuccessful. Steam automobiles never caught on; they were simply not practical. Attempts by Samuel P. Langley (1834–1906), head of the Smithsonian Institution, to build a steam-powered airplane ended in failure. His efforts were soon overshadowed by the brothers Orville Wright (1871–1948) and Wilbur Wright (1867–1912), who used a more versatile form of power to lift the world's first self-propelled airplane off the ground: the gasoline engine.

A contemporary diesel engine, displayed in France in 2006.

GASOLINE AND DIESEL ENGINES

Without the gasoline engine, the Wright brothers could never have taken to the air. Yet this crucial piece of technology was still quite new when the Wright brothers fastened it to their glider in 1903. First conceived in the seventeenth century, gasoline engines appeared in the 1860s. In 1864, German engineer Nicolaus August Otto (1832–1891) built the first practical engine. This machine burned gasoline made from petroleum to push pistons and turn wheels.

Gasoline engines were significantly more efficient than engines powered by steam. They could get more energy from the same weight of fuel and were consequently far lighter and more compact than steam engines. These characteristics made gasoline engines perfect for use in airplanes. Gasoline was also cleaner, creating less smoke pollution, and made refilling engines convenient.

All these advantages opened the door for gas-powered road vehicles, developed in the 1870s by two other German engineers, Karl Benz (1844–1929) and Gottlieb Daimler (1834–1900). Their inventions inspired another German, Rudolf Diesel (1858–1913), to invent a diesel engine in 1890. This engine was designed to burn almost any kind of fuel, even oil made from peanuts and waste animal fat, though all diesel engines burned a low-grade petroleum oil that came to be known as diesel.

ELECTRICITY

Gasoline and diesel were ideal forms of energy for transportation, but in homes and factories another kind of energy, elec-tricity, had an equally significant impact. Electricity is a natural phenomenon, so it was discovered rather than invented. Nevertheless, a whole series of scientists and inventors played a part in turning it from a scientific novelty into a usable form of energy.

Static electricity, the kind that builds up in one place, was known to the ancient Greeks and later explored by such scientists as Benjamin Franklin (1706–1790). It was of little use until others figured out how to turn it into current electricity, which can flow from place to place. The first to do so were Italian anatomy professor Luigi Galvani (1737–1798) and his friend Alessandro Volta (1745–1827), a professor of physics. Galvani thought electricity was

The Etiwanda Generating Plant near Ranch Cucamonga, California.

Power to the People

Simple tools and early inventions such as the wheel gave individuals the power to improve their lives, but the engines that came later had a very different impact. Steam engines were huge and costly machines that only the rich could afford and that concentrated power in the hands of wealthy industrialists.

Gasoline engines have proved popular because they give many people access to power. Only a few people could afford to run steam engines at the height of their popularity in the 19th century, but in the 21st century many can afford cars and trucks. Gasoline engines, however, make the world dependent on a steady supply of oil. This was one reason that Rudolf Diesel invented his diesel engine, which could run on almost any fuel: he originally imagined people growing their own "energy crops," so local communities could become entirely self-sufficient. In the future, vehicles may be driven by solar panels or other energy sources.

Since the end of the 19th century, most electricity has been generated in huge power plants and distributed over power lines to homes, offices, and factories some distance away. This process concentrates power in the hands of utility companies and is also very wasteful. Around two-thirds of energy made by power plants is used in the plants themselves or in the wires that carry electricity to consumers. When all the energy losses are considered, only 22 percent of the energy generated in power plants is used productively.

Just as gasoline and diesel engines shifted power away from huge, inefficient steam engines, so too may renewable forms of energy have a similar effect on electricity. Local communities can now build their own small power plants that generate heat and electricity closer to where they are used. This process prevents wasting energy in power lines. Individual homes, schools, factories, and businesses can install solar cells and small wind turbines. These have the potential to turn any building into a tiny power plant that makes its own inexpensive energy. Apart from dramatically cutting electricity costs and heating bills, such technologies reduce pollution, cut fossil fuel consumption, and help to address problems like global warming. Scientists believe technologies like this could cut $2.7 trillion from the world's energy bill by 2030.

made inside the bodies of animals. While trying to disprove this theory in the late 18th century, Volta discovered he could make electricity flow by connecting two different metals through a chemical liquid. He invented the battery, the first practical source of electricity, in 1800.

Volta's method of turning chemicals into electricity was one influence on Michael Faraday (1791–1867), an English scientist who came to appreciate the close relationship between electricity and magnetism. In 1821, this led him to develop the first crude electric motor, a device that used electricity to make a wheel rotate. Ten years later, he invented the exact opposite of a motor: the electric generator, which produced electricity when magnets or coils of wire spun inside it.

Faraday's work hinted at the exciting potential of electricity. Another half century or so was to pass, however, before Thomas A. Edison (1847–1931) realized that potential when he developed the practical electric lightbulb in the 1870s. Edison then built the world's first electric power plants in the following decade. Once homes had been wired for electricity, other inventors developed various electric appliances in the early 20th century. New inventions and reworkings of older inventions that use electricity continue to be brought to market by creative minds.

An example is British inventor James Dyson (1947–), who reconfigured the vacuum cleaner in the 1980s and 1990s to require no dust bag.

Electricity is the most modern and convenient form of energy: it can be generated at a power plant in one place and then carried many miles along wires, across state borders, to the homes, offices, and factories where it is eventually consumed. Inside power plants, the generators (loosely based on those developed by Faraday and Edison) spin at high speeds, but they, too, require power to function. In most plants, the generators are turbines, machines like windmills that rotate when high-pressure steam blows past them. The steam is made by boiled water, which can be heated in various ways. Many power plants boil water by burning coal, oil, or natural gas; others use nuclear power.

NUCLEAR ENERGY

Early in the 20th century, German-born physicist Albert Einstein (1879–1955) developed a remarkable new way of understanding the world—the theory of relativity, which postulated that matter and energy were really the same. Einstein's most famous equation, $E = mc^2$, expresses

TIME LINE

ca. 3500 BCE	27 BCE – 395 CE	ca. 600 CE	ca. 1710	1792 – 1800	1800
Wheel developed in ancient Mesopotamia.	The Romans pioneer the use of treadmills.	Windmills are developed in the Middle East.	Thomas Newcomen develops the steam engine.	Alessandro Volta and Luigi Galvani discover current electricity.	Volta invents the battery.

enabled other scientists to see that when large atoms are split, energy is released. If atoms were arranged to keep splitting indefinitely in what is known as a chain reaction, a colossal amount of energy would be released. If the reaction were uncontrolled, the result would be a nuclear explosion.

During World War II, Fermi worked as part of the Manhattan Project, a top-secret, $2 billion effort to develop nuclear weapons for the United States. In 1942, he developed the first nuclear reactor, an apparatus inside which atoms are split to produce energy. Three years later, his work was used to develop the nuclear bombs that were dropped on Japan in August 1945. In the years that followed, tensions mounted between the United States and the Soviet Union (the former alliance of Russia and its republics) and the two superpowers battled to develop the greatest arsenals of nuclear weapons. During the 1950s, American scientists led by Edward Teller (1908–2003) developed an

the concept that even a tiny amount of matter can produce an enormous amount of energy. Italian physicist Enrico Fermi (1901–1954) understood the implications of Einstein's amazing ideas. Fermi's work

TIME LINE

1814	1821	1831	1864	1870s	1890
George Stephenson develops first practical steam locomotive.	Michael Faraday invents the first electric motor.	Faraday invents the electric generator.	Nicolaus August Otto constructs the first practical engine.	Karl Benz and Gottlieb Daimler develop gas-powered road vehicles; Thomas Edison develops the first practical electric lightbulb.	Rudolf Diesel creates the diesel engine.

even more powerful weapon—the hydrogen bomb—that was capable of an awesomely destructive release of energy. The hydrogen bomb has not yet been used in human wars.

RENEWABLE ENERGY

Although engines that run on gasoline and diesel can be considered among the most successful inventions ever developed, they have adverse effects on the environment, as do many electric plants powered by coal, oil, and gas. These substances are known as "fossil fuels" because they were formed from plants and animals that decomposed underground over millions of years. Fossil fuels proved to be inexpensive, convenient, and abundant in the 20th century, but they are slowly beginning to run out. Burning fossil fuels in engines or power plants also causes pollution and reportedly contributes to global warming, a gradual rising of earth's temperature that is slowly changing the climate.

Scientists, engineers, and politicians are increasingly turning their attention to forms of renewable energy. These are ways of making power that do not rely on finite supplies of fuels, that produce little or no pollution, and that can operate indefinitely. Although renewable energy is top-ical, it is far from a new idea. The Greeks and Romans who developed waterwheels were pioneers of clean, renewable energy in ancient times. Numerous inventors have rediscovered water power more recently. One was American engineer Lester Pelton (1829–1918), who explored ways of using creeks to help power mining equipment during the 19th-century California gold rush. He designed an improved overshot waterwheel with cups around its edge to extract more energy from slowly running water. Pelton's waterwheel was later used in hydroelectric power plants, which produced electricity using the energy in rivers.

Since all of earth's energy ultimately comes from the sun, solar energy is thought by many to offer the most promising long-term answer to the planet's energy needs. Among his many other achievements, Albert Einstein was the first scientist to figure out how light can be used to generate electricity. His theory, published in 1905, was known as the photoelectric effect. A number of pioneering inventors have since worked to develop solar technology. American physicist Stanford Ovshinsky (1922–) helped to promote the use of solar cells, which make electricity directly from sunlight. Maria Telkes (1900–1995) built the first solar-powered house in the 1930s.

TIME LINE (continued)

ca. 1860	1903	1930s	1942	1950s	20th century
Lester Pelton develops and refines a waterwheel later used in hydroelectric power plants.	The Wright brothers make the first flight.	Maria Telkes builds the first solar-powered house.	Enrico Fermi develops the first nuclear reactor.	Edward Teller heads the development of the hydrogen bomb.	Stanford Ovshinsky develops and promotes solar cells.

Where Does Our Energy Come From?

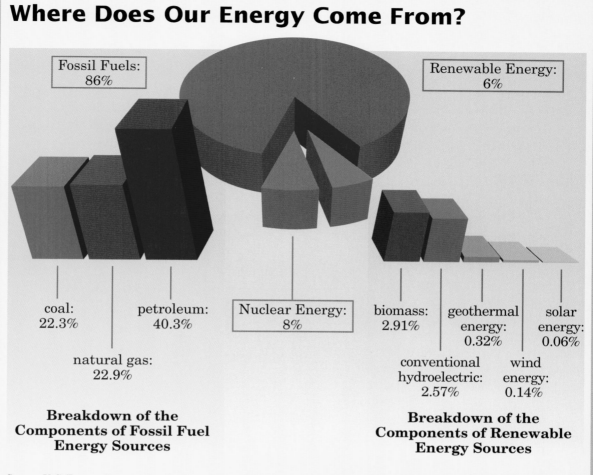

Fossil Fuels:
86%

Renewable Energy:
6%

coal:
22.3%

petroleum:
40.3%

natural gas:
22.9%

Nuclear Energy:
8%

biomass:
2.91%

geothermal
energy:
0.32%

solar
energy:
0.06%

conventional
hydroelectric:
2.57%

wind
energy:
0.14%

**Breakdown of the
Components of Fossil Fuel
Energy Sources**

**Breakdown of the
Components of Renewable
Energy Sources**

Source: U. S. Energy Consumption by Energy Source, Energy Information Administration, *Annual Energy Review 2005.*
< http://www.eia.doe.gov/emeu/aer/pdf/pages/sec1_9.pdf> (October 13, 2006).

*Data from the Energy Information Administration
shows the nation's energy consumption by source.*

As earth's population continues to grow, supplies of fossil fuels continue to dwindle. Renewable energy offers the best hope of meeting everyone's needs for energy. Despite the pressing need for renewable technologies, some inventions remain relatively expensive and inconvenient. Many years may pass before most people drive electric cars or generate their own power from the sun and wind. Inventors will con- tinue to play a pivotal role in shifting society from its dependence on fuels like oil and coal to a future where energy is clean, affordable, and endlessly available.

—Chris Woodford

Further Reading

Books

Challoner, Jack. *Eyewitness: Energy.* New York: Dorling Kindersley, 2000.

Ramage, Janet. *Energy: A Guidebook*. Oxford and New York: Oxford University Press, 1997.

Woodford, Chris. *Energy and Electricity*. New York: Facts On File, 2004.

Web sites

Center for Alternative Technology
A comprehensive introduction to renewable energy.
http://www.cat.org.uk/

Energy Information Administration: Kids Page
Information, games, and activities from the U.S. Department of Energy.
http://www.eia.doe.gov/kids/

Energy Quest: Energy Education
An introduction from the California Energy Commission.
http://www.energyquest.ca.gov/

See also: Benz, Karl; Brunel, Isambard Kingdom; Daimler, Gottlieb; Diesel, Rudolf; Dyson, James; Edison, Thomas; Faraday, Michael; Fermi, Enrico; Franklin, Benjamin; Fulton, Robert; Newcomen, Thomas; Otto, Nicolaus August; Ovshinsky, Stanford; Pelton, Lester; Stephenson, George; Telkes, Maria; Teller, Edward; Volta, Alessandro; Wright, Orville, and Wilbur Wright.

DOUGLAS ENGELBART

Inventor of the computer mouse
1925–

Computers became widespread in the second half of the 20th century for two main reasons. First, technical improvements made computers smaller, more powerful, and less expensive. Second, computers became considerably easier to use. By the end of the 1970s, these once complex machines had found their way into homes, schools, and businesses. Nothing short of a revolution, this was largely the work of pioneers such as Douglas Engelbart, an American computer scientist who developed many ways of making computers more user-friendly. His most famous invention is the computer mouse.

EARLY YEARS

Douglas Carl Engelbart was born on January 20, 1925, in Portland, Oregon. His grandparents had been pioneer settlers. His childhood was spent on a farm in the harsh years of the Great Depression, milking cows and looking after chickens. In 1942, he graduated from high school and then entered Oregon State University, where he was an honors student in his senior year and earned a bachelor of science degree in electrical engineering.

Like many others, Engelbart found his studies interrupted by World War II. From 1944 through 1946, he served in the navy in the Philippines. With his background in electrical engineering, he made an ideal electronic technician, helping to maintain radar (a radio device that ships and airplanes used to navigate) and sonar (a navigation device, based on sound waves, used on ships and submarines). Engelbart realized, as he watched how radar operators used their equipment—sitting in front of giant television-like cathode-ray tube (CRT) displays—that he was witnessing something very important about the links between humans and the complex machines they were operating.

THE CRUSADE BEGINS

When the war ended, Engelbart returned to his studies, graduating in 1948. He then moved to the San Francisco area and took a job with the National Advisory Committee for Aeronautics, the forerunner of the National Aeronautics and Space Administration (NASA). Around 1950, at age 25, he had a sudden moment of panic. As he recalled later, he had a good job and he was about to get married—and he wondered, What was there left to achieve? As he contemplated what to do with the rest of his life, he had a vision of the future: "I got an image of myself sitting at a big

Douglas Engelbart in 1997, with a pet mouse sitting on top of the first computer mouse.

CRT screen with all kinds of symbols on it, new and different ones, manipulated by a computer that could be operated through various input devices." There and then he began to form what he later called his "crusade": to redesign computer-based equipment in a way that would truly improve human lives.

He quit his job the following year and enrolled as a graduate student at the University of California at Berkeley. Over the next few years, he studied for his doctorate in electrical engineering; the work he did in this period enabled him to apply for 19 patents on electronic equipment. By 1955, he was an assistant professor, full of what he called "wild ideas" for using new technology. One of his colleagues quietly advised that, if he kept on promoting these ideas, he would stay an assistant professor forever. Engelbart decided he should relocate to where his creativity and vision would be appreciated. Stanford University and the Hewlett-Packard Company both turned down his request to work in computing because, at that time, they felt computers had little future.

HUMANS AND COMPUTERS

In 1957, Engelbart joined Stanford Research Institute (SRI). Within a few years, he had his own group, which he later named the Augmentation Research Center. The group's unusual name came from a pioneering proposal, Augmenting Human Intellect, that Engelbart wrote in 1962. In it, he imagined creative people such as architects sitting at CRT displays and manipulating what they could see on the screen with "a small keyboard and various other devices." The idea was to use computer technology to extend (augment) what people could do with their brains (human intellect).

During the 1960s, Engelbart and his growing team (eventually 47 people) set about turning this vision into reality. Beginning in 1962, they developed a user-friendly computer system called NLS (short for oNLine System), which pioneered most of the easy-to-use technologies that people now take for granted. They filed a patent for the most famous of these—the computer mouse—in 1967; it was granted in November 1971 (see box, The Mouse).

On December 9, 1968, Engelbart and 17 of his colleagues attended the Fall Joint Computer Conference in San Francisco and gave a 90-minute demonstration of their work. Around one thousand of the world's leading computer scientists were present in the hall, and they sat in silence, astonished, as Engelbart showed off a series of groundbreaking innovations. Apart from the mouse, the group presented videoconferencing (the ability to talk to colleagues by means of a computer screen), on-screen word processing (at that time, all documents were edited on paper), collaborative working (in which several people could use their mice to control a single computer screen), hypertext (a way of linking documents together, later made popular by the World Wide

The Mouse

Early computers had no screens, keyboards, or mice. At that time computers were expensive and complex, and operating them was a very skilled and time-consuming job. When Douglas Engelbart worked as a radar technician, he realized that people instinctively looked at things, pointed at them, and moved them with their hands. Engelbart wanted computers to be that easy to use, too.

The computer mouse is the most famous result of Engelbart's work. It was originally called an "X-Y position indicator for a display system." However, this technical jargon was quickly forgotten when someone noticed that the cable hanging from the front of the device looked like a mouse's tail. The original mouse had a wooden case and two wheels inside that could roll along at right angles. As the operator moved the mouse, either or both wheels turned to a small degree. An electronic circuit inside measured the wheel movements and sent this information down the cable to the computer. A program running in the computer calculated how much to move the cursor on the screen.

Douglas Engelbart remembers making his first designs for the mouse in 1961. It has changed relatively little over the intervening years. Some mice still work mechanically, like the original. They typically have a heavy rubber ball underneath that turns two small wheels hidden inside the plastic case. Newer mice use an optical system instead. A bright light-emitting diode (LED) shines red light down from the mouse onto the desk. The light reflects back into a light-detecting component (photocell). As the mouse moves, the pattern of reflected light changes. Using this information, an electronic circuit inside the mouse calculates how to move the cursor on screen. In the future, computers may have built-in scanning devices that recognize people's eye movements. Instead of pointing with a mouse, the user will simply look at part of the screen and the cursor will move straight there.

A 21st-century wireless mouse.

Web), windows (multiple documents open on the same screen at one time), and a "chordal" keyboard that let people type any letter with the fingers of just one hand. Some years later, Engelbart vividly recalled the audience reaction: "I looked up and everyone was standing, cheering like crazy."

MIXED FORTUNES

Of all these inventions, Engelbart was able to patent only the mouse: the others were mostly software (programming) innovations, and lawyers had deemed they could not be protected under patent law. His ideas were soon being adapted and developed elsewhere. In the 1970s, members of the SRI team went to work at the Xerox Company's Palo Alto Research Center, which incorporated some of Engelbart's ideas into a $40,000 workstation—the Alto. Although never a commercial success, this machine inspired Steve Jobs, one of the founders of Apple Computer, to develop his easy-to-use Macintosh in the early 1980s. The Macintosh, in turn, strongly influenced the "look and feel" of Microsoft Windows, the world's most-used computer program, in the 1990s. Thus, Douglas Engelbart's work helped to make several generations of computer systems easier to use from the 1970s onward.

For Engelbart himself, however, things turned out less happily. He received no royalties from inventing the mouse, and even SRI made little money from the invention. When the company licensed it to Apple, SRI was paid just $40,000 for a device that now sits on virtually every desk in the world. Perhaps the main reason for this failure to profit was that the mouse was about 20 years ahead of its time when it was first developed. Engelbart's original patent expired in 1977, several years before the mouse really caught on as a computer control device.

In 1977, SRI sold NLS and Engelbart's laboratory to another company, Tymshare, Inc., and funding was cut sharply. In the mid-1980s, Tymshare was bought out by the giant McDonnell Douglas Corporation, which closed down Engelbart's laboratory altogether in 1989. It was one

TIME LINE

1925	1944	1955	1957	1971	1989	2000
Douglas Carl Engelbart born in Portland, Oregon.	Engelbart serves in the U.S. Navy as an electronic technician.	Engelbart receives doctorate in electrical engineering.	Engelbart joins the Stanford Research Institute.	Engelbart's team receives patent for computer mouse.	Engelbart founds the Bootstrap Institute.	Engelbart awarded National Medal of Technology.

of the more eventful years of Engelbart's life: the same year, his house burned down while he and his family stood outside, watching helplessly. Undeterred by these setbacks, in 1989, Engelbart and his daughter Christina founded the Bootstrap Institute to promote his ideas.

RECOGNITION AT LAST

When the Apple Macintosh and Microsoft Windows became popular in the mid-1980s, people finally started to recognize Engelbart as a true pioneer of user-friendly computing. From the late 1980s, he received many honors and awards. In 1998, he was inducted into the National Inventors Hall of Fame. Two years later, President Bill Clinton awarded him the National Medal of Technology for "creating the foundations of personal computing." The computer world needed more than two decades to realize the usefulness of Engelbart's most famous invention, the mouse. It may take just as long for the rest of his ideas to become popular.

—Chris Woodford

Children playing with the first prototype of the Xerox Alto computer; the Alto incorporated many of Engelbart's ideas and later served as inspiration for the Apple Macintosh.

Further Reading

Books

Bardini, Thierry. *Bootstrapping: Douglas Engelbart, Coevolution, and the Origins of Personal Computing.* Stanford, CA: Stanford University Press, 2000.

Williams, Brian. *Computers: Great Inventions.* Portsmouth, NH: Heinemann, 2001.

Wurster, Christian. *Computers: An Illustrated History.* New York: Taschen, 2002.

Web sites

Bootstrap Institute
　　Douglas Engelbart's own Web site.
　　http://www.bootstrap.org/

Howstuffworks: How computer mice work
　　An illustrated guide to the workings of mechanical and optical mice.
　　http://computer.howstuffworks.com/mouse.htm

See also: Computers; Jobs, Steve, and Steve Wozniak.

ENTERTAINMENT

Many inventions make life easier. Tractors make it quicker for fewer people to farm the food that societies need, and computers have automated routine clerical jobs. One result of inventions like these is that individuals have more time in which to entertain themselves. In addition to helping to create leisure time, inventors have also devised many ways for people to fill such time—with inventions such as musical instruments, photography, television, and movies for personal and family entertainment.

MUSIC

Music making is one of the oldest forms of human entertainment. Historians think people have been making sounds with drums since at least the early Stone Age (around 600,000 years ago), when the first drums were created using hollowed-out logs. These simple percussion instruments are the oldest musical inventions. Drums made from skins, wind instruments, and string instruments are much more recent. Archaeologists believe flutes made from hollowed-out animal bones were in use around

A woman playing a lyre on a fourth-century Christian sarcophagus found in Rome.

10,000 BCE, and lyres (simple, handheld harps) became popular after 3000 BCE. Many instruments are even newer than this. Violins are believed to have been invented in the early sixteenth century. Valved brass instruments date from the 1800s.

The piano has been among the most popular musical instruments for home entertainment since it first appeared in the early 18th century. Italian instrument maker Bartolomeo Cristofori (1655–1731) invented the piano while he was trying to improve the harpsichord, an older instrument. Like a piano, a harpsichord has a keyboard; when the keys are pressed, levers inside the case pluck a series of strings. These levers always pluck the strings with the same force, thus preventing the player from making the notes louder or softer to suit different types of music. One solution was the invention of harpsichords with two keyboards: one that played loud notes, the other soft. Cristofori had a better idea: he replaced the levers with small wooden hammers, creating an instrument called the *gravicembalo col piano e forte*, or "harpsichord that could play loud or soft." The name was soon shortened to pianoforte and then piano.

The piano was the height of musical technology in the 18th century, and musicians have continued to develop new inventions ever since. In the 19th century, for example, scientist Charles Wheatstone (1802–1875) began his career by developing a whole series of musical instruments, the most famous of which was the concertina (a handheld version of the accordion). Another 19th-century inventor, Adolphe Sax (1814–1894), devised a

number of new brass instruments, including the saxotromba and the saxophone (patented in 1846). More recent musical inventions include the solid-bodied electric guitar, developed in the 1940s and 1950s by Les Paul (1915–).

In modern times, people play music in their homes and cars using technology such as CDs (compact discs), which were invented in the late 1960s by James T. Russell (1931–). Until the late 19th century, the only way to hear music was to play it at home on an instrument or to attend a live concert. Hearing recorded music first became possible when the great American inventor Thomas Edison (1847–1931) developed his phonograph in 1877. This machine recorded sounds by digging a groove into a sheet of foil, which was wrapped around a metal cylinder, as the operator cranked a handle. Although primitive by modern standards, it worked,

An iPod from 2006 that can not only play music but can also show television shows and movies, and play games.

and the basic technology remained in use for more than a century. A similar concept was used in the gramophone, which evolved from Edison's phonograph and played music from a revolving disc. In 1948 gramophone records in turn evolved into long-playing (LP) records that rotated at $33\frac{1}{3}$ revolutions per minute (rpm). Invented by American engineer Peter Goldmark (1906–1977), LPs could store about half an hour of music on each side of a vinyl (plastic) disc.

Since the 1980s, people have started listening to music in an entirely different way: using headphones and compact music players as they move about in their daily lives. This first became possible when the head of the Japanese Sony Corporation, Akio Morita (1921–1999), inspired his company to develop a pocket-size portable cassette player, which was named the Walkman. The portable cassette player later evolved into the portable CD player, the mini-disc player, and the MP3 player.

PHOTOGRAPHS

Humans have been drawing pictures for just about as long as they have been making music—since prehistoric times. The earliest preserved images are petrograms (drawings made on the walls of caves), made in the Stone Age and earlier. Until the 18th cen-tury, drawing and painting were the only ways to make permanent records of images from everyday life. The first to experiment with photographs were English scientists Thomas Wedgwood (1771–1805) and Humphry Davy (1778–1829), who used paper coated with chemicals to record crude images. The chemicals, based on silver, were sensitive to light, changing their structure when exposed to light and making black or white patches on the coated paper.

French inventor Joseph-Nicéphore Niépce (1765–1833) made the first photograph in 1827, using silver plates coated with silver iodide, but his images were not permanent. When exposed to the light, the plates turned black and the images were lost. Over the next few years, Niépce worked closely with French artist Louis Daguerre (1789–1851) to improve his invention. After Niépce's death, Daguerre discovered a way of making detailed, permanent images on metal plates. He called this invention the daguerreotype, and such prints were very popular in the United States during the second half of the 19th century. The year Daguerre announced his invention, Englishman William Henry Fox Talbot (1800–1877) unveiled an alternative method of making photographs. He began by using his camera to make a reversed image called a negative, in which the black areas appear white and vice versa. After removing the negative from the camera, he could make any number of final prints, or positives, by treating the paper with chemicals. Although Daguerre's method was initially more popular, Talbot's process superseded it, and film-based printing still uses essentially the same negative-positive process.

Photography was a form of entertainment, but in its early days it was also considered a science. Pioneers like Daguerre and

Talbot were chemists as much as they were photographers: every photograph they took had to be developed (turned into a final print) through a careful, multistage process using bottles and trays of chemicals. Taking photographs was difficult until American George Eastman (1854–1932) invented his simple, handheld Kodak cameras. His first major innovation was plastic photographic film, which he patented in 1884. His easy-to-use Kodak box camera appeared four years later, followed by his immensely popular Kodak Brownie camera in 1900. Eastman's genius was to make photography simple enough for anyone. He spurred the creation of a huge industry and helped to turn photography into a popular hobby during the 20th century.

MOVIES

Photography was a way of recording history, but only as individual moments of time captured in single, frozen images. From the middle of the 19th century, inventors began to experiment with ways of turning still photographs into moving images. In 1861, American inventor Coleman Sellers (1827–1907) developed a machine he called the kinematoscope. He mounted still photos on a rapidly rotating paddle wheel so that, as the wheel spun, the photos merged in the viewer's eye to make a moving image.

Making movies was not simply a matter of displaying many individual photographs to a viewer at high speed. Such photographs had to be taken by a camera that could capture many still images in quick succession. Considering that Niépce's original camera had needed eight hours to make a single image, this was no small feat. By the 1880s, French scientist Étienne Jules Marey (1830–1904) had developed a camera that could record a series of images onto a piece of moving film. In the following decade, Thomas Edison and his assistant William Dickson (1860–1937) made their kinetograph, the world's first practical movie camera. Edison and Dickson's invention was capable of capturing 48 photographs each second onto a piece of film that rattled through the camera when the operator cranked a handle. Once the images had been filmed, they could be viewed with another piece of apparatus, a kinetoscope. This was a bit like a giant microscope into which the viewer had to peer; only one person could view at a time.

Modern movies were born when the French brothers Auguste and Louis Lumière (1862–1954 and 1864–1948) saw Edison's invention and resolved to develop something better. In 1895, they patented their cinématographe, a single device built into a wooden box that could take a series of photographs, develop them, and project them onto a wall for whole groups to watch. The Lumières started shooting short movies of everyday life and opened the world's first movie theater in 1896. From then on, moviegoing became an immensely popular form of entertainment. Hollywood began to dominate movie production during the 1920s, when sound also was introduced. The first color movies appeared in the 1930s. By that time, however, another new form of visual entertainment—television—was being developed.

RADIO AND TELEVISION

Television combined elements of radio and the movies. From radio, television borrowed the idea of sending information over long distances without wires; movies supplied the idea of filming images in one place and playing them back for an audience elsewhere.

Radio is a means of sending information from one place to another using electromagnetic waves: invisible patterns of electricity and magnetism that travel at the speed of light (186,000 miles per second). During the 19th century, European scientists figured out what radio waves were and how they could travel. In 1901, Italian inventor Guglielmo Marconi (1874–1937) demonstrated how radio could send messages across entire oceans, and radio moved from the science laboratory into the wider world. Although initially used to improve communication between ships, radio became a popular form of entertainment after 1910, when the first public radio broadcast was made from the Metropolitan Opera in New York City.

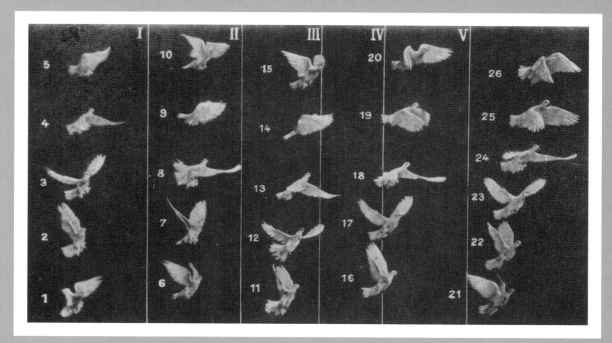

An early example of motion photography by Étienne Jules Marey shows a bird in flight.

Fads

Hula hoops, Frisbees, Rollerblades, and skateboards—fads like these are usually taken up faster than any other kind of invention. Although they may be trivial in comparison with great inventions such as steam engines and penicillin, they nevertheless bring pleasure to millions of people.

Fads have been sweeping the world since mass-produced goods first became popular in the late 19th and early 20th centuries. Even though the first Kodak cameras cost $25 when they went on sale in 1888 (equivalent to more than $500 today), they sold by the thousand. Mass production was the secret behind many other fads. On May 15, 1940, the day nylon stockings first went on sale, customers swarmed to buy almost five million pairs at around $1.25 per pair (equivalent to more than $17 today).

The first Frisbees were made from empty metal tins in which the Frisbie Pie Company sold its wares. Students at Yale University reputedly ate the pies, then threw the empty tins to one another for entertainment. During the 1950s, a company called Wham-O changed the name slightly and started manufacturing plastic Frisbee discs. Wham-O's next success came with the hula hoop. Hula is a traditional dance in Hawaii, and it became a worldwide craze in 1958 when Wham-O started selling large plastic hoops. Within months, millions of children and teenagers were amusing themselves by wobbling the hoops around their hips. One more recent fad is the Super Soaker water gun, invented in the 1980s by an aerospace engineer, Lonnie Johnson (1949–).

Some fads that achieve lasting popularity evolve into sports. Skateboards are thought to have been invented in the early 1960s when people started attaching roller skates to wooden boards. Modern, aggressive skateboarding began as a fad in Los Angeles, California, about a decade later. In-line skates date back to the 18th century, but became popular again in the 1980s when brothers Scott and Brennan Olson (1959– and 1963–) developed Rollerblades.

Invented in 1958, the hula hoop was very loosely based on the traditional Hawaiian dance, the hula.

The drawback of radio was that it could send and receive only sound, not pictures. As the inventors of movies had found, making a moving picture required building a machine that could capture many still images. Instead of taking photographs, however, the inventors of television had to find a way of capturing images in electrical form to be transmitted by radio waves—they had to develop television cameras.

Three inventors successfully tackled this problem in the 1920s. A Scots inventor, John Logie Baird (1888–1946), made one of the first television cameras using a large wooden disc with a spiral pattern of holes in it. As the disc rotated, the holes "scanned" whatever was being filmed using a light-detecting unit placed behind it: each hole in the disc captured a part of the image like a rapidly blinking eye. More advanced cameras were developed around the same time by Russian-born American physicist Vladimir Zworykin (1889–1982) and American radio engineer Philo T. Farnsworth (1906–1971).

Early television cameras could detect only patterns of light and dark, so the first televisions made black-and-white pictures. Color television appeared in 1940, invented by Peter Goldmark, the American who went on to develop LP records. During the following decade, Charles Ginsburg (1920–1992) of the Ampex Corporation developed the video-tape recorder, enabling ordinary people to store and play back their favorite television programs.

MODERN ENTERTAINMENT

Music, photography, television, and movies are still widely enjoyed in the 21st century, and many more traditional forms of entertainment remain just as popular as ever. Fairground attractions, for example, have been entertaining millions since they were first introduced in the 19th century. The first Ferris wheel was built by American engineer George Ferris (1859–1896) for the world's fair in Chicago in 1893; similar wheels are still being constructed and enjoyed today. The modern roller coaster was developed by John A. Miller (1872–1941) in the early 20th century. Historians believe the idea may have originated in Russia, perhaps five to six hundred years ago, when people rode blocks of solid ice down tall icy wooden frameworks. Using a much more elaborate system of wheels and tracks, Miller enabled roller coasters to go higher, steeper, and faster than ever before.

TIME LINE

ca. 10000 BCE	3000 BCE	Early 1600s	ca. 1700	1800s	1827
Evidence of flutes made from hollowed-out animal bones.	Lyres become popular.	The violin is invented.	Bartolomeo Cristofori invents the piano.	Valved brass instruments invented.	Joseph-Nicéphore Niépce makes the first photograph.

The midway and rides at the Minnesota state fair in 2005.

Inventors are continually developing new ways for people to amuse themselves as technology advances. The arrival of the microchip (single-chip computer) in the early 1970s made possible many new forms of entertainment. Electronic games based on microchips became popular thanks to American inventor Nolan Bushnell (1943–), who launched the Atari Corporation in 1972. Atari's first major game, PONG, was an electronic version of Ping-Pong that could be played on a home television. Games like these were one of the inspirations for affordable home computers, which were popularized by Steve Jobs (1955–) and Steve Wozniak (1950–), two Californians who set up the Apple Computer Corporation

TIME LINE (continued)

1839	1846	1877	1884	1888	1893
Daguerre perfects the daguerreotype; Talbot introduces photographic paper.	Adolphe Sax patents the saxophone.	Thomas A. Edison develops the phonograph.	George Eastman patents plastic photographic film.	Eastman begins to sell his Kodak camera.	George Ferris builds the first Ferris wheel.

in 1976. More recently, personal computers (PCs) have allowed people to entertain themselves using the Internet, a network that allows computers to exchange information almost instantaneously. When Tim Berners-Lee (1955–   ) created the Internet-based World Wide Web in 1989, he invented a way for people to find information, meet new friends, shop online, and pursue all kinds of other hobbies without even leaving their homes.

Computer technology has also enabled some of the more traditional forms of entertainment to be reinvented. Photography, long based on using chemical films, has now evolved into digital photography, in which images are stored in computer files. Photographs taken with digital cameras can be uploaded onto Web sites or e-mailed to friends and family members. Music has also been reinvented by digital technology. Instruments such as electronic synthesizers—pioneered in the 1950s by Harry Olson (1901–1982) of the RCA Corporation—can mimic the sound of anything from a piano to an entire orchestra. In much the same way, television and radio have moved into the digital age, with programs now broadcast in digital form (signals are transmitted as strings of coded numbers). This results in better sound and picture quality and more channels. It also enables viewers and listeners to "interact" with programs by, for example, taking part in opinion polls, answering quiz questions, or selecting favorite camera angles during baseball games.

If history is any guide, even better technologies will be developed. Consequently, people may spend even less time working and even more on leisure pursuits. Inventing itself may become a more popular pastime. The development of the Internet, for example, has made collaboration on social and cultural projects between two or more individuals increasingly popular. One of these collaborations is Wikipedia, a large online encyclopedia written collectively by tens of thousands of people; another example is Linux, a computer operating system invented by Finnish programmer Linus Torvalds (1969–   ) and now written by thousands of programmers, many of them amateurs. The word amateur originally meant someone who loved what he or she did; it reflects how the difference between work and entertainment disappears when people can do what they enjoy. One challenge for inventors of the future is to develop more creative forms of work that people find truly entertaining and rewarding.

—Chris Woodford

TIME LINE

1896	1900	1910	1920s	1948	Late 1960s
Auguste and Louis Lumière open the world's first movie theater.	Eastman introduces the Kodak Brownie camera.	First public radio broadcast made from New York City.	Baird, Zworykin, and Farnsworth independently develop television cameras.	Peter Goldmark invents long-playing (LP) records.	James T. Russell invents the compact disc.

Attendees at the 2006 Electronic Entertainment Expo test out Playstation 3 game consoles.

Further Reading

Books

Ardley, Neil. *Eyewitness: Music*. New York: Dorling Kindersley, 2004.

Buckingham, Alan. *Eyewitness: Photography*. New York: Dorling Kindersley, 2004.

Florida, Richard. *The Rise of the Creative Class: And How It's Transforming Work, Leisure, Community, and Everyday Life*. New York: Basic Books, 2004.

Rosenblum, Naomi. *A World History of Photography*. New York: Abbeville, 1997.

Web sites

Bad Fads Museum
    The history of popular fads.
    http://www.badfads.com/

Songs in the Key of E
    A history of electronic music from the IEEE Virtual Museum.
    http://www.ieee-virtual-museum.org/

See also: Berners-Lee, Tim; Bushnell, Nolan; Cristofori, Bartolomeo; Daguerre, Louis; Eastman, George; Edison, Thomas; Ferris, George; Ginsburg, Charles; Goldmark, Peter; Jobs, Steve, and Steve Wozniak; Johnson, Lonnie; Lumière, Auguste, and Louis Lumière; Marconi, Guglielmo; Miller, John A.; Morita, Akio; Olson, Brennan, and Scott Olson; Paul, Les; Russell, James T.; Sax, Adolphe; Torvalds, Linus; Wheatstone, Charles.

TIME LINE (continued)

1972	1976	1980s	1989	2006
Nolan Bushnell launches the Atari Corporation.	Steve Jobs and Steve Wozniak start Apple Computer Corporation.	Sony Corporation introduces the Walkman.	Tim Berners-Lee creates the World Wide Web.	Sony releases its Playstation 3 game console.

ENVIRONMENT AND INVENTING

In 1861 American writer Henry David Thoreau (1817–1862) noted in his journal: "Thank God, men cannot as yet fly, and lay waste the sky as well as the earth!" Since then, people have not only taken to the sky—in both airplanes and rockets—but also developed cars, electricity, plastics, pesticides, chemical plants, atomic bombs, air conditioners, aerosols, and many more things that harm the earth. Fortunately, modern inventors are also devising new technologies that are helping to arrest and reverse some of the environmental damage done by their predecessors.

INDUSTRY AND URBANIZATION

During the 18th and 19th centuries, a series of inventions dramatically changed the western world in a period of history known as the Industrial Revolution. It began when 18th-century inventors such as Englishman James Hargreaves (ca. 1720–1778) devised machines that could process cloth more quickly. Soon, machines were making goods faster and less expensively than humans could. In the United States, Eli Whitney (1765–1825) is famous for inventing the cotton gin (a machine for cleaning harvested cotton plants), but he also developed the idea of mass production (in which unskilled people make many identical goods in a facto-

ry). Just over a century later, Henry Ford (1863–1947) invented the assembly line, where workers mass-produced inexpensive cars that moved past them on a conveyor belt. Around 81,000 people worked at Ford's biggest factory in River Rouge, Michigan. In Hargreaves's time, no one could have imagined human industry on such a scale.

Machines, mass production, and assembly lines—inventions like these helped to urbanize much of the world as people moved from rural areas to work in factories (and later in offices). In 2001, the American Association for the Advancement of Science (AAAS) made headlines when it published a series of dramatic satellite photographs revealing that humans have paved or plowed roughly a quarter of the planet, and that they have somehow changed approximately half its land surface area. In total, half the world's people now live in urban areas; in developed countries, the figure is more than three-quarters. Even in sub-Saharan Africa, one of the world's least developed regions, more than 40 percent of the people dwell in towns and cities.

Urbanization and industrialization have undoubtedly brought many benefits to society. People are now wealthier and more highly educated than ever before; they are also generally healthier and live longer. Yet the shift from a rural to an

Shifting Populations: Urban vs. Rural Growth Rates, 1950 to 2030 (Projected)

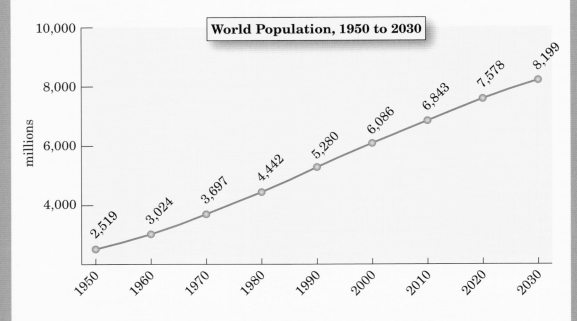

World Population, 1950 to 2030

millions

10,000 — 8,000 — 6,000 — 4,000

2,519 · 3,024 · 3,697 · 4,442 · 5,280 · 6,086 · 6,843 · 7,578 · 8,199

1950 1960 1970 1980 1990 2000 2010 2020 2030

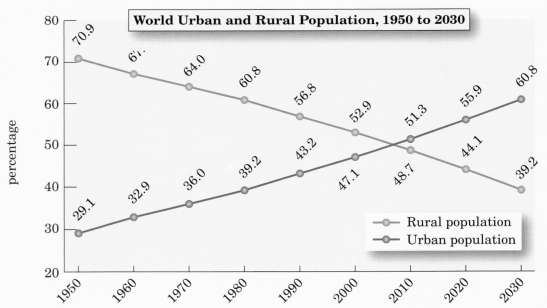

World Urban and Rural Population, 1950 to 2030

percentage

80 — 70 — 60 — 50 — 40 — 30 — 20

Rural population: 70.9 · 67.1 · 64.0 · 60.8 · 56.8 · 52.9 · 51.3 · 55.9 · 60.8

Urban population: 29.1 · 32.9 · 36.0 · 39.2 · 43.2 · 47.1 · 48.7 · 44.1 · 39.2

1950 1960 1970 1980 1990 2000 2010 2020 2030

Rural population
Urban population

Source: Population Division of the Department of Economic and Social Affairs of the United Nations Secretariat, *World Population Prospects: The 2004 Revision and World Urbanization Prospects: The 2003 Revision*, http://esa.un.org/unpp, accessed October 11, 2006.

If historical trends hold steady over the coming years, urban growth and rural decline will continue and the majority of the world's population will live in urban areas.

urban way of life has also brought problems. Towns and cities need vast supplies of energy, create pollution and waste, and consume earth's finite resources faster than they can be replenished. Environmentalists believe society cannot continue in this way and argue that human life on earth is becoming unsustainable.

FARMS AND FACTORIES

During the Industrial Revolution, as towns and cities grew, farming was transformed. Inventions such as the polished steel plow devised by John Deere (1804–1886) made farming easier and greatly expanded both the amount of land used for growing crops and its productivity. From the 1830s, Cyrus Hall McCormick (1809–1884) developed a series of reaping machines that could harvest crops more easily. Deere and McCormick's machines were pulled by horses—and therefore were limited by how long and hard horses could work. However, at the start of the 20th century Henry Ford

produced his affordable Fordson tractor, which could pull heavier loads than horses and for as long as it had fuel.

Machines like these led to the production of more and more food by fewer and fewer workers in the 19th and 20th centuries. In 1800, before the Industrial Revolution had properly begun in the United States, around 90 percent of people worked on farms; 200 years later, fewer than 2 percent farm the land. As people moved from farms into factories, agriculture became an industry too, and farms became factories of a different kind.

However, "intensive agriculture," as this is known, came at a cost. Fields did not produce all this food without assistance—a vast input of energy (used to fuel tractors), artificial fertilizers (chemicals that restore nutrients to the soil that growing crops remove), and pesticides. Over the last fifty years or so, worldwide fertilizer use has increased more than tenfold. A quarter of the world's agricultural land that has been irrigated (watered) for crop growth has been damaged in the process.

Water flows into an irrigation ditch beside a cornfield in Arizona.

A great deal of water is wasted by watering crops inefficiently, which can have disastrous consequences. Around fifty countries (and 35 percent of the world's population) are currently experiencing water shortages that place millions of lives—especially those of young children—at grave risk.

ENGINES AND MACHINES

Power for the Industrial Revolution came with the invention of the steam engine. An engine is a machine that makes heat energy by burning fuel. In a steam engine, burning coal boils water and generates steam, which pushes pistons and turns wheels. Invented in 1712 by Thomas Newcomen (1663–1729), steam engines required far greater amounts of coal to be mined from the ground. The burning coal also produced smoke that gradually began to choke towns and cities, and waste ash that had to be cleared away. Steam engines powered factories, but they also found their way onto ships and locomotives. In the 19th century, steamships and trains opened up countries and continents, allowing people to travel farther than ever before and encouraging the urbanization of even more rural land.

About 150 years after the invention of the steam engine, the gasoline engine was developed by German engineer Nicolaus August Otto (1832–1891). Gasoline

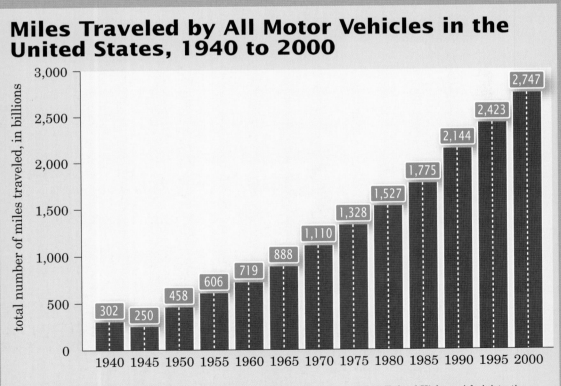

Miles Traveled by All Motor Vehicles in the United States, 1940 to 2000

Source: Data from Highway Statistics 2001 and *Highway Statistics Summary to 1995*. Federal Highway Administration. U. S. Department of Transportation. http://www.fhwa.dot.gov/policy/ohpi/hss/hsspubs.htm, accessed October 11, 2006.

Miles of road traveled in the United States have increased every year since World War II ended in 1945.

engines found their way first into automobiles and motorcycles, pioneered by Karl Benz (1844–1929) and Gottlieb Daimler (1834–1900), and later into airplanes, developed by Orville Wright (1871–1948) and Wilbur Wright (1867–1912). Like steam engines before them, gasoline engines hugely increased the need for fuel (this time, petroleum).

Although gasoline engines produce less pollution than steam engines, there are many more of them—for example, an estimated 500 million gas-powered cars in the world—so the overall effect on large cities has proven to be much the same or worse. In the United States, which has 500 cars for every 1,000 residents, the American Lung Association has estimated that half the country's urban areas now regularly exceed pollution limits and more than 132 million people are at risk of developing lung problems such as asthma and bronchitis.

CHEMICALS AND MATERIALS

During World War II, Swiss chemist Paul Müller (1899–1965) saved many soldiers' lives by developing DDT (dichlorodiphenyltrichloroethane), a chemical pesticide (insect killer). After the war, farmers embraced DDT and many other pesticides and herbicides (weed killers), spraying them in huge quantities. Only in the 1960s did the toxicity and persistent (long-lasting in the environment) nature of these chemicals become widely known, largely because of the work of American ecologist Rachel Carson (1907–1964). In her famous book of 1962, *Silent Spring*, she wrote: "As crude a weapon as the cave man's club, the chemical barrage has been hurled against the fabric of life." After Carson's book created a worldwide panic about toxic pollution, many of the most harmful agricultural chemicals were banned.

Yet chemicals are now more popular than ever. By the 1990s, the U.S. Environmental Protection Agency estimated that more than one hundred thousand different chemicals were on the market, including nearly nine thousand food additives, seventy-five hundred cosmetic products, and six hundred different pesticides—with around two thousand new chemicals being introduced each year. One reason for this immense growth in the production of different kinds of chemicals is that many modern materials are entirely synthetic (not derived from plants, animals, or other natural sources). In 1907, Belgian-born chemist Leo Baekeland (1863–1944) began a materials revolution when he invented Bakelite, the world's first plastic. During the 1930s, American chemist Wallace Carothers (1896–1937) produced an even more versatile plastic, nylon. Plastics have since transformed the world, finding their way into everything from stockings and toothbrushes to Tupperware food containers, marketed from the 1940s on by Earl Tupper (1907–1983).

Plastics were inexpensive and heralded a new era, when manufacturers started making disposable products and people began to throw away objects and products they no longer liked or needed without even thinking about it. Ironically, most disposable products do not break down naturally: discarded plastic fishing line can remain in the oceans for six hundred years; disposable diapers and plastic bottles can take more than four hundred years to break down in landfills. Toxic waste is another "invention" of the industrial age. The world's factories now produce close to

Around two thousand new chemicals are introduced every year.

half a billion tons of hazardous waste every year; polar bears in the remotest parts of the Arctic have been found to contain traces of toxic chemicals that originated in factories thousands of miles away.

GLOBAL WARMING

If scientists are to be believed, the price of progress will have to be paid by future generations in the form of global warming, a gradual rise in the temperature of earth's atmosphere since the Industrial Revolution. Global warming is caused by a buildup of gases such as carbon dioxide in the atmosphere, causing the world to heat up like a greenhouse. Most of these "greenhouse gases" are a by-product of the burning of fuel by engines and machines, though some come from natural processes as well.

Scientists estimate that, mostly as a result of human activities, earth's atmosphere now has more carbon dioxide than at any time during the last 400,000 years. With such a huge transformation of the atmosphere, global warming is believed to be causing climate change, with more extreme and erratic weather around the world. Another projected consequence is a rise in sea levels—up to three feet— during the 21st century, with catastrophic effects on low-lying islands and coastal areas around the world. Climate change could also lead to famines, water shortages, and difficulties in producing enough food.

Opinions are divided about how to tackle the problem. One idea is that people should drastically reduce the carbon dioxide they produce. This could mean insulating homes to reduce waste heat, using renewable forms of energy such as solar and wind power, switching to energy-efficient household appliances, and using mass transportation instead of personal automobiles whenever possible. Another solution is to try to reduce the carbon dioxide in the atmosphere. Plants use carbon dioxide from the air in a chemical reaction called photosynthesis. Thus planting trees is one way to tackle global warming. Many people are now exploring an idea called carbon offsetting: calculating how much carbon dioxide they produce each year (driving their car, heating their home, and so on) and then paying a

A solar-powered traffic light in front of the Tiananmen Gate in Beijing, China.

conservation group to plant some trees to compensate.

A high-tech version of the same idea is called carbon sequestration; it involves removing carbon dioxide produced by engines and machines (for example, using "scrubbers" on smokestacks) and then pumping it into underground reservoirs for long-term storage. A study by the United Nations has shown that half of all global carbon dioxide emissions might be stored using this technology by 2050.

If, as seems likely, climate change is now unstoppable, inventors will have to help populations to adapt to a warmer world. For example, people will need new and stronger materials to build coastal defenses against rising sea levels. More desalination plants (which make fresh-water from seawater) will also be needed. Diseases like malaria could spread more widely, so improved medicines and vaccines will also be important.

REINVENTING THE ENVIRONMENT

Human inventiveness, which has cured or ameliorated terrible diseases and fired rockets to the moon, may yet find solutions to the problems now threatening earth's environment. Inventors never deliberately set out to make toxic chemicals, engines that cause pollution, fertilizers that damage fields and groundwater, or plastics that wash up on beaches. Yet some inventors have purposely focused their attention on developing technologies that address these problems.

The Ozone Hole

The ozone layer is earth's natural sunscreen: a region of the atmosphere about 12–30 miles (19–48 km) above sea level that naturally blocks some of the sun's harmful ultraviolet rays. As the name suggests, it consists of ozone, a type of oxygen gas.

In 1928, American chemist Thomas Midgley Jr. (1889–1944), invented chemicals called chlorofluorocarbons (CFCs) that seemed to be perfect for use in aerosols, refrigerators, and air conditioners. Because they are not toxic and do not readily react with other substances, CFCs were long thought to be harmless. Then, in 1974, two American chemists predicted that CFCs would cause a thinning out (depletion) of the ozone layer. The prediction was verified in 1985, when other scientists found a huge hole, as big as the United States, in the ozone layer over Antarctica. Another huge hole, this time over the Arctic, was discovered in the 1990s. After studying the ozone problem, the U.S. Environmental Protection Agency predicted that the loss of ozone could lead to an extra 154 million cases of skin cancer and 3.4 million deaths.

Against this background, many nations banned CFCs under an international agreement, the Montreal Protocol. Industry started using other, less harmful gases. For example, one of the pioneers of air conditioning, the Carrier Company—named for its founder Willis Carrier (1876–1950)—was also the first to produce a domestic air conditioner that used chemicals that were not harmful to ozone. However, with the damage to the ozone layer already done, another solution was needed to protect people from the sun. Hence the increased use of sunscreens, which contain chemicals that either absorb or reflect the sun's ultraviolet light, stopping it from reaching people's skin.

Yet even this "technological solution" may have brought problems of its own. Sunscreens contain a cocktail of thirty to forty different chemicals and, when people shower after going to the beach, the chemicals drain into the sewage system. Washing into rivers and seas, they are thought to disrupt endocrines (hormones) in the bodies of fish and other marine creatures, causing changes in sex and other problems. Scientists have now invented soy-based sun creams that avoid the more harmful chemicals. Initial tests suggest these are more effective at blocking out the sun, but present no risk to fish when they wash into waterways.

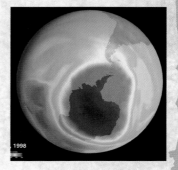

NASA graphic shows the distribution of ozone in Dobson units, which are the basic measurement used in ozone research. The hole in the ozone layer over Antarctica is represented in purple.

Climate change may make more desalination plants necessary, like this one in Palma de Mallorca, Spain.

Solar power (energy from the sun's radiation) may be one way to combat earth's energy crisis. Stanford Ovshinsky (1922–) pioneered solar cells, which turn sunlight directly into electricity, and Maria Telkes (1900–1995) designed the world's first solar home in the 1930s. Others have tried to develop different forms of renewable energy. Windmills have been used as a source of power since at least 600 CE, although the world's first modern wind turbine was built in Rutland, Vermont, in the 1940s. Another renewable energy source, water power, has been used to drive machines since ancient times. Modern hydropower dates to the 19th century, when American engineer Lester Pelton (1829–1918) developed an improved water wheel to make power during the California gold rush.

For many environmental problems, technological solutions are now at hand. For example, six-pack rings used to package drink cans are sometimes made from a photodegradable plastic (one slowly destroyed by light) that naturally disintegrates in about six months. Recycling initiatives help to reduce the amount of solid waste (for example, glass, paper, and cardboard) in landfills. Emissions from various forms of transportation have one of the greatest impacts on earth; however, ways to reduce the problem have been found and implemented. In California, for example, tax incentives encourage people to drive electric vehicles, which not only help combat problems such as smog pollution but are also more energy-efficient. Globally, communications technologies such as telephones, fax machines, and the Internet also have the potential to reduce the need for routine travel, cutting both toxic emissions and the energy people consume.

Although technological fixes can solve some problems, no one knows whether they will create new environmental problems of their own (see box, The Ozone Hole). Many environmentalists believe placing faith in technology is unwise. They think people should learn to live "sustainably" within earth's natural limits, as earlier societies did. Others disagree with this view. They point out the tremendous material gains that have been made through technological progress and feel that society cannot return to simpler times. There is little disagreement that humans have changed the earth dramatically; the question is how long people can continue to exploit the planet before threatening their own survival.

—Chris Woodford

Further Reading

Books

Burnie, David. *Endangered Planet*. Boston: Kingfisher, 2004.

Kudlinski, Kathleen. *Rachel Carson: Pioneer of Ecology*. New York: Puffin, 1989.

Leggett, Jeremy. *The Carbon War*. New York: Penguin, 2000.

Levine, Shar, and Alison Grafton. *Projects for a Healthy Planet: Simple Environmental Experiments for Kids*. San Francisco: Jossey-Bass, 1992.

Steingraber, Sandra. *Living Downstream*. New York: Vintage, 1998.

Web sites

Kids for Saving Earth
   Environmental education to help children save the planet.
   http://www.kidsforsavingearth.org/

U.S. Environmental Protection Agency: Kids page
   Information and activities about all aspects of the environment.
   http://www.epa.gov/kids/

See also: Baekeland, Leo; Benz, Karl; Carothers, Wallace; Carrier, Willis; Daimler, Gottlieb; Deere, John; Ford, Henry; Hargreaves, James; McCormick, Cyrus Hall; Newcomen, Thomas; Otto, Nicolaus August; Ovshinsky, Stanford; Pelton, Lester; Telkes, Maria; Tupper, Earl; Whitney, Eli; Wright, Orville, and Wilbur Wright.

JOHN ERICSSON

Marine engineer and inventor of
a screw propeller

1803–1889

Between 1800 and 1900, warships evolved from wooden vessels, which were powered by sails with cannons mounted on wheeled carriages; to iron vessels, which were powered by steam engines and featured fixed guns in revolving turrets. Perhaps the single most important marine engineer involved in bringing about this transition was John Ericsson, who over a career of more than fifty years developed a wide range of new and improved inventions for ships, including engines, propellers, and weapons.

EARLY YEARS

Johan Ericsson (Ericsson's birth name) was born in Långbanshyttan, in the province of Värmland, Sweden, on July 31, 1803. His father Olof was a works manager for a mine and later for the Göta Canal. Johan was trained as a surveyor on the canal, at that time one of the biggest civil engineering projects in Europe. His elder brother Nils later became a famous canal engineer in Sweden. After service in the Swedish Army, however, Johan decided to immigrate to Britain, where he hoped to find an outlet for his growing skills as a mechanical engineer.

Ericsson's interest in steam engines led him to partner with engineer John Braithwaite. In early 1829 they installed a steam engine in the 85-ton (77-metric ton) ship *Victory*, in which John Ross (1777–1856), a famous Scots explorer, led his four-year expedition to the Arctic. The machinery featured paddles that could be lifted out of the water to avoid being damaged by ice, and a condensing boiler installed below the waterline for greater stability. In a condensing boiler, water from the condensed steam was returned to the boiler for reuse. The builders' inexperience showed, however; the boilers and pumps were faulty from the start, never working for more than a few hours at a time before leaking or breaking down. Ross eventually had the engine removed and abandoned it in the Arctic, relying on sail thereafter.

A scale model of Ericsson and Braithwaite's locomotive, the Novelty.

In October 1829 Ericsson and Braithwaite entered the Rainhill Trials, a competition to choose a locomotive for the Liverpool and Manchester Railway, the world's first all-steam railroad line. On the first day of the trials, Ericsson and Braithwaite's locomotive, the Novelty, was the fastest of all the competitors, reaching a top speed of 28 miles per hour (45 km/h). Overheating damaged the boiler, however, and the Novelty could not be repaired in time to continue the remaining runs. That left only the Rocket, built by the British inventor George Stephenson (1781–1848) and his son Robert (1803–1859), still running; the Rocket accordingly won the trial.

THE SCREW PROPELLER

In July 1836 Ericsson registered a British patent for a screw propeller for use in steamships. By that time, many people had already had the idea that a screw, placed at the back of a ship and powered by a steam engine, would be a better way of propelling a ship through the water than paddle wheels, especially for warships. Unlike paddles, a screw would be below the waterline and thus harder to attack. A screw would also be more stable than paddles, which in rough seas caused a ship to zigzag slightly as each paddle lifted alternately out of the water.

Undated oil painting of John Ericsson by C. L. Elliot.

Ericsson's important innovation was to use two propellers mounted on coaxial shafts (one shaft inside the other) and turning in opposite directions to give his system both greater power and greater stability.

The following year Ericsson built a steamboat, the *Francis B. Ogden*, propelled by this system, and demonstrated it to the heads of the British navy. The navy did not buy the rights to Ericsson's design, noting correctly that with the propeller mounted behind the rudder, the boat could not be steered effectively.

At this point Ericsson met Robert Stockton (1795–1866), a U.S. Navy captain who was willing to use his personal wealth to fund construction of a new vessel featuring Ericsson's propeller. The ship, named the *Robert F. Stockton*, was built near Liverpool and set sail for the United States in April 1839. Ericsson himself followed later that year. Stockton used his connections in government (his father was a U.S. senator) to get the U.S. Navy to build a screw-propelled warship designed by Ericsson. With the propeller now moved in front of the rudder, the 700-ton (635-metric ton) corvette USS *Princeton* was launched in 1843. Shortly after the launching, however, a large gun designed by Stockton exploded as it was being test-fired, killing several prominent politicians on board at the time. Stockton managed to shift blame for the disaster onto Ericsson, although the engineer had nothing to do with the gun's design or construction. Stockton also ensured that because of the explosion, Ericsson did not get paid for his work on the *Princeton*.

Ericsson then devised a twin-shaft propeller system, in which two separate propellers were evenly spaced on either side of a ship's centerline. An improvement on a single, central propeller shaft for wooden ships, the twin-shaft propeller avoided cutting through the central sternpost, which risked weakening the ship's structure. During the rest of the 1840s and the 1850s Ericsson's patented propeller designs steadily spread across the United States, where they were used on river and lake craft as well as on oceangoing ships. In 1848 Ericsson became a naturalized American citizen.

A NEW DESIGN

Soon after the beginning of the Civil War in 1861, Ericsson was persuaded to return to work for the U.S. Navy, despite his experience with the USS *Princeton*. The navy knew that the Confederacy was building an ironclad warship (formerly the USS *Merrimack*, relaunched as the CSS *Virginia*), and the Union needed an iron ship of its own. Ericsson came up with an extraordinary design based on one he had proposed to the French emperor, Napoleon III, in 1854. It used an idea pioneered in 18th-century forts, in which the fort's thick defensive walls were set in deep ditches below ground level, shielding the fort from cannon fire aboveground. Ericsson followed this idea by designing his ship, named the *Monitor*, with almost all of its bulk beneath the waterline. Its flat deck was just 18 inches (45 cm) above the waves, giving an enemy

TIME LINE

1803	1829	1836	1837	1843	1862	1881	1889
Johan Ericsson born in Långbanshyttan, Värmland, Sweden.	Ericsson and John Braithwaite install steam engine in the *Victory*.	Ericsson registers a British patent for a screw propeller.	Ericsson builds the *Francis B. Ogden*.	Fatal explosion aboard the USS *Princeton*.	The *Monitor* confronts the *Merrimack*.	The *Destroyer*, a torpedo boat, first launched in New York City.	Ericsson dies.

almost nothing to fire at except a single revolving turret that housed the ship's large guns.

The *Monitor*'s guns and turret, designed by Ericsson, were great innovations. The British designer Cowper Coles had already designed a revolving turret that ran on rails. Ericsson's turret allowed the big 11-inch (28-cm) guns, which had to be loaded from the front, to be pulled back from the turret openings quickly and safely for reloading, then moved forward again for firing. Previously such guns had been on wheeled carriages, which had to be hauled back and forth slowly by large teams of men pulling on ropes.

The *Monitor* met its rival, the *Merrimack*, in Hampton Roads on March 9, 1862; the two iron ships fired their cannon at each other for four hours, sometimes at almost point-blank range, though the shells simply bounced off the heavily armored hulls of both ships, inflicting no structural damage. The engagement proved that the *Monitor*'s design was a success, and many more ships like it were built, mainly to support troop action in bombarding coastal targets from rivers or close to shore. Ericsson took his design to his native Sweden, which built many such ships for its own navy. The design was then imitated and adapted by the navies of Europe over the next two decades.

THE TORPEDO

After the Civil War, Ericsson became interested in the torpedo. Originally designed for coastal defense, the torpedo found a new application in the world of iron warships that Ericsson himself had helped to create. The Croatian engineer Ivan Blaz Lupis Vukic (1813–1875) had

Powered by Hot Air

An idea that fascinated Ericsson for a long time was the heat engine. First proposed by Scots clergyman Robert Stirling in 1816, the heat engine would heat air that was fed into a piston directly, eliminating the need for steam. Such an engine would be more efficient than a steam engine (or an internal combustion engine) and also less dangerous. While still a Swedish army officer in the early 1820s, Ericsson made a small heat engine powered by burning birch wood. In 1833 he made a larger version, which he called a "caloric" engine (from the Latin word for "heat"), that would run on coal. This 5-horsepower engine was followed by a 24-horsepower version in 1838.

After moving to New York, Ericsson built further experimental versions during the 1840s; in 1852 he persuaded a group of businessmen to build a ship powered by an enormous caloric engine. Named the *Ericsson* after its designer, it was 260 feet (79 m) long and propelled by paddles, and was launched the following year. Although the engine, with its huge cylinders 14 feet (4.25 m) in diameter, worked as planned, its movement was too slow, and the ship never developed enough speed to be commercially viable for carrying freight or passengers. After the ship sank in shallow water during a storm, it was salvaged and refloated, and Ericsson removed the caloric engine and replaced it, regretfully, with a steam engine.

demonstrated the first clockwork-powered surface torpedo to the Austrian emperor in 1860; with the British engineer Robert Whitehead (1823–1905), Vukic also developed a version, powered by compressed air, that ran underwater.

In 1873 Ericsson developed a torpedo that featured his trademark pair of contrarotating propellers on a single shaft, with the torpedo suspended from floats at a fixed depth underwater. It was powered by an electric motor that received its charge from wires connected to the ship or land station from which the torpedo was fired. The wires spooled out behind the torpedo as it moved forward. The wires also enabled the rudder to be controlled, making the torpedo one of the first weapons that could be guided after firing.

It quickly became clear that to threaten battleships, torpedoes would need to be fired from small, fast vessels able to operate in the open sea. The torpedo boat that Ericsson designed for this purpose, the

Destroyer, was launched in New York City in 1881. In this boat Ericsson further developed the idea, used in the *Monitor*, of protecting a vessel by placing most of its bulk underwater. Although it could not dive, the *Destroyer* was almost submarine-like: tapered at both ends, with

Screw propellers continue to power ships in modern times; here, workers inspect the newly installed screw propeller on the USS Kitty Hawk in 2003.

completely watertight decks and able to cruise with just its shielded wheelhouse above water. Also like later submarines, the ship fired its torpedoes from a tube below the waterline.

Unlike Ericsson's 1873 model, however, the torpedo fired from the *Destroyer* had no power of its own: it was effectively a huge bullet, and the firing tube was the barrel of a gun that propelled the torpedo by a gunpowder charge. Consequently the torpedo had a range of only about 100 yards (90 m—about one-tenth the range of self-propelling torpedoes of the time), but a very high speed of about 60 knots (70 mph or 110 km/h). Unlike the *Monitor*, the *Destroyer* never saw action in wartime, and its torpedo system was soon superseded by further improvements to self-propelled torpedoes, whose greater range made them much safer for an attacker to use.

Ericsson's inventions covered an amazingly wide range; his patents, particularly on the screw propellers, made him a wealthy man. By the time of his death in New York in 1889, Ericsson was widely considered to be among the greatest mechanical engineers of his time. A year after his death, his body was returned to Sweden and interred in an impressive mausoleum in his native province of Värmland.

—Jonathan Dore

Further Reading

Books

Church, William Conant. *The Life of John Ericsson*. Honolulu, HI: University Press of the Pacific, 2003. (First published in 1890.)

DeKay, James Tertius. Monitor: *The Story of the Legendary Civil War Ironclad and the Man Whose Invention Changed the Course of History*. New York: Walker, 1997.

Thulesius, Olav. *The Man Who Made the* Monitor: *A Biography of John Ericsson, Naval Engineer*. Jefferson, NC: McFarland, 2006.

Web site

Monitor Center

Extensive resource page about the *Monitor*, including its history, design, and the preservation of its salvaged remains.

http://www.monitorcenter.org

See also: Fulton, Robert; Stephenson, George; Transportation.

CARRIE EVERSON

Inventor of the ore concentration process
1842–1914

Carrie Everson's pioneering work in mining technology paved the way for engineers to extract more precious metals from silver, copper, and gold ore. Although mining engineers later recognized the importance of her work and used her method of extracting metals from ore, she received no benefit from her discoveries. It was only after her death that she was credited as having contributed significantly to the mining industry.

EARLY YEARS

Everson was born Rebecca Carrie Jane Billings near Sharon, Massachusetts, on August 27, 1842. She received an excellent education; in 1864, she married William Knight Everson, a Chicago doctor. Her husband encouraged her to continue her education, and she studied medicine and chemistry whenever she had the time.

Sometime in the 1860s or 1870s, Everson's husband began buying stock in Colorado mining companies; this prompted her to extend her studies to the subject of mineralogy. Bad decisions were made, however. In 1878, Mark M. "Brick" Pomeroy, a gold promoter, persuaded Everson's husband to invest his entire fortune of $40,000 in a Colorado gold mine. The mine did not exist and the Eversons was plunged into poverty.

Silver was discovered in Colorado in 1878, and, as more and more silver deposits were found in that state, thousands of people moved there to make their fortune. The Eversons settled for a time in Georgetown, Colorado. A silver boomtown, Georgetown was home to inventors, mining engineers, metallurgists, and many others at the forefront of silver mining technology. While her husband was on a trip to Mexico, Everson conducted experiments in the field of metallurgy, the science and technology of metals, in the hope of making a profitable discovery. She began investigating ways to improve the method of extracting silver, gold, and copper from ore.

THE FLOTATION PROCESS

Traditional mining methods relied on gravity to separate the heavier metals from the lighter sand and quartz found in ore. Everson discovered, however, that if she first mixed pulverized ore and oil together and then added acid and water to the mixture, the metal particles adhered, or stuck, to the oil that floated to the top. The metal could then be collected more easily than by traditional gravity

An undated portrait of Carrie Jane Billings Everson.

methods. Everson's process eventually became known as the flotation process. Not only was it a more effective way to extract metal from ore, but the flotation process also made possible the extracting of metal from low-grade ore—which contains less metal than high-grade ore and which previously had been ignored, as the cost to extract the ore was too costly and the process too inefficient.

Everson conducted dozens of experiments to determine the most effective oils. She tested petroleum, lard, cottonseed oil, castor oil, oil from the sperm whale, and linseed oil, and discovered that metal would adhere to all of them. When applying for her patent, however, she recommended the use of cottonseed oil because it was the least costly. She also conducted experiments using a variety of acids, including hydrochloric acid, sulfuric acid, nitric acid, acetic acid, and oxalic acid. As was the case with the many oils she tested, all acids were effective; she recommended the use of sulfuric acid, again because it was the least costly.

Everson and her husband engaged a patent attorney in Denver to assist them in registering Everson's new process with the U.S. Patent Office in Washington, D.C. In 1886, Everson's patent, named "Process of Concentrating Ores," was registered in her name.

For the next two years, Everson and her husband tried to interest Colorado mining companies in her oil-flotation process. Yet no mining engineers were interested enough to persuade their companies to invest. In the late 1880s, Colorado was still in the heyday of its silver boom. Ore rich in silver deposits was so plentiful that mining companies had no need to invest in a method that would enable them to get metal from low-grade ore.

In January 1889, Everson's husband died, and she lost her best promoter. With the encouragement of her friends, she found a new business partner, Thomas Criley. He advertised the Everson process throughout Colorado, and then in the Pacific Northwest. Just as he was close to striking a deal with mine owners in Oregon, he, too, died. Everson rushed westward to complete the deal but was unable to make a sale.

A miner holds up a rock with traces of gold ore.

Everson's next partner was Charles Hebron, a New York chemist. Mining historians consider their project, a revised process of recovering ore that they patented in 1892, to be less efficient than Everson's 1886 process. The partnership failed, and Everson did not profit from the patent. After this disappointment, Everson stopped trying to profit from her discoveries. She concentrated on her work as a visiting nurse for the Denver Flower Mission, a job she held for 17 years. Everson died in 1914.

TIME LINE

1842	1878	1886	1889	1905	1914
Everson is born near Sharon, Massachusetts.	Everson and family move to Colorado.	Everson receives patent for flotation process.	Everson's husband dies.	A British company builds a mill in Australia using Everson's method.	Everson dies.

REDISCOVERING EVERSON

In 1905, a British mining company in Australia built the first mill that used Everson's oil-flotation method. However, historians believe that this company knew nothing of Everson's work and developed the method on

Copper and gold are separated from other minerals in a flotation machine in the Philippines in 2006.

its own. Then, in 1912, James Hyde, a mining engineer in Montana, developed an oil-flotation method that was used in copper mining. In 1914, when British and American mining companies realized that they were both using the same method, a conflict ensued in which a number of companies claimed the right to patent the flotation process. When the lawyers for these companies searched the U.S. patent archives, they found Carrie Everson's 1886 patent.

As an inventor, Carrie Everson was extraordinary because she was a woman inventor in metallurgy and mining, two fields dominated by men. Even in the 21st century, mining historians do not agree on the importance of Everson's discoveries to the development of the mining industry, in large part because it is unknown whether the mining engineers who developed the process in later years had any knowledge of her discoveries or her patent. Nevertheless, Carrie Everson is recognized as having been the first person to patent the oil-flotation method.

—Judith Harper

Further Reading

Book
Altman, Linda Jacobs. *Women Inventors*. New York: Facts On File, 1997.

See also: Buildings and Materials.

DANIEL FAHRENHEIT

Inventor of the mercury thermometer
and temperature scale
1686–1736

The Fahrenheit scale for measuring temperature was devised in the 18th century by Daniel Fahrenheit. Fahrenheit also introduced the first mercury thermometer and was the first to demonstrate how boiling-point temperatures vary with atmospheric pressure.

EARLY YEARS

Daniel Gabriel Fahrenheit was born into a well-to-do household in Danzig (modern-day Gdansk, Poland) on May 14, 1686, the eldest of five children. His father was a merchant. Both his parents died on the same day in 1701; it is thought that they ate poisonous mushrooms. His four younger siblings were placed in foster homes, but Daniel Fahrenheit, then 15, was apprenticed to a merchant in Amsterdam, the Netherlands, to train as a bookkeeper. There Fahrenheit became interested in scientific instruments. After his apprenticeship, he abandoned bookkeeping and worked as a glassblower.

Interested in expanding his knowledge of scientific instrumentation, Fahrenheit began to travel sometime in 1707. He wandered extensively across northern Europe, seeking out scientists and instrument makers in Denmark, Poland, Germany, and Sweden, as well as in England.

THE MERCURY THERMOMETER

Many scientists throughout the seventeenth century and the early 18th century experimented with instruments capable of measuring temperature. The great expansion of scientific inquiry in this period led many to devise such tools as part of a larger interest in measuring and quantifying all natural phenomena. Variations among the many types of thermometers of the time, however, made it impossible to refer to a common measure of air temperature.

During the 1720s, Fahrenheit had been constructing and working with different kinds of thermometers. Those of the time normally used a combination of water and alcohol and were better than the gas (air) thermometers that had been in use since the early 1600s. However, they were still inaccurate (see box, History of the Thermometer).

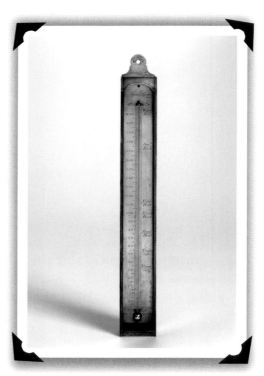

One of the scientists Fahrenheit met on his travels was the Danish astronomer Olaus (or Ole) Roemer. Roemer had invented a thermometer that used alcohol to measure temperature. Alcohol was preferred to water because,

A mercury-in-glass thermometer made by George Adams around 1780; the temperature scale ranges from minus 13 to 217 degrees Fahrenheit.

when sealed inside the glass, it was less susceptible to changes in atmospheric pressure. By 1714, Fahrenheit had constructed his first thermometers. That year, he demonstrated two of them: both contained alcohol, and they gave identical air temperatures—the first time such a feat had been accomplished.

Also in 1714, Fahrenheit devised the first mercury thermometer. He used mercury because it expanded at a much more constant and predictable rate than either alcohol or water and could be used to measure a wider range of temperatures. Sealed inside newly improved glass tubing, Fahrenheit's mercury remained unaffected by changes in air pressure. With this significantly more stable instrument for measuring and recording the temperature of substances, Fahrenheit set about creating a standard scale that could be used by people in different places.

History of the Thermometer

Although Fahrenheit is considered the inventor of the modern thermometer, instruments for measuring heat or cold, in fact, go back many years earlier. Galileo Galilei (1564–1642) was one of the first to develop an instrument to measure the expansion or contraction of air inside a glass container.

Around 1600, Galileo poured water through an opening in a long, narrow glass tube connected to a glass bulb. As the air under the water expanded (warmed) the water would rise; and as it contracted (cooled), the water level would lower. Water, however, did not provide very accurate readings, because it was so easily affected by changes in atmospheric pressure. Called a thermoscope, Galileo's instrument had no means of measuring temperature, so it was not a proper thermometer. In 1612, an Italian inventor, Santorio Santorio (1561–1636), applied a scale to a glass tube similar to the one Galileo used, thereby allowing the measurement and recording of temperature.

Galileo's thermoscope.

DEVELOPS TEMPERATURE SCALE

At the time, many different techniques and scales were applied for making basic observations about air temperature. Comparing the temperature in Amsterdam with the temperature in London was impossible without a common method of calculation. After meeting with Roemer in Denmark, Fahrenheit saw the need to base his system on two fixed points.

Roemer had been the first to introduce this idea; he had chosen as his upper point the boiling point of water (which he labeled 60) and as his lower point the melting point of ice (which he labeled 7.5). What Roemer had not realized was that the boiling point of water varied with atmospheric pressure—a relationship that Fahrenheit was the first to comprehend. The effect of pressure on boiling points is observable on a mountaintop, where the pressure is lower than at sea level; water at higher altitudes boils at a lower temperature.

Wanting to base his system on more commonly occurring temperatures, Fahrenheit chose the temperature of the human body as his upper point (which he originally set at 96) and the temperature of a mixture of salt, water, and ice (0 degrees) as his lower point. He then divided his system into 96 equal parts and established the freezing point of water at 32 degrees. Later, the boiling point of water was calibrated to be 212 degrees (see box, Fahrenheit's Temperature Scale).

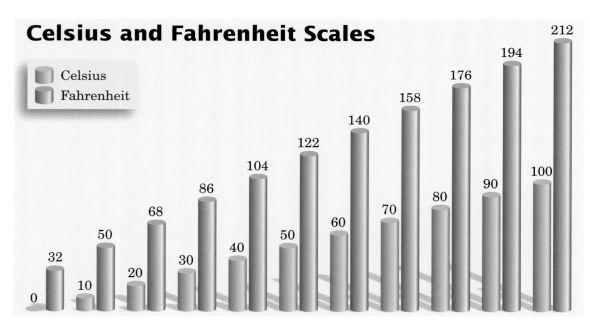

Two main temperature scales are used throughout the world today: the Fahrenheit scale and the Celsius scale, which was proposed in 1742 by the Swedish physicist Anders Celsius (1701-1744). The Celsius scale is used in most countries outside the United States. The temperature in Celsius is sometimes referred to as centigrade, which means "divided into one hundred units." A mathematical formula is used to convert from one scale to the other: $C = (5/9)(F – 32)$, where C is the temperature in Celsius and F is the temperature in Fahrenheit.

Fahrenheit's Temperature Scale

After Daniel Fahrenheit invented a mercury thermometer capable of more consistent temperature readings, he established a scale that could be used everywhere so that people could refer to a common number for a given temperature—including the boiling and freezing points of water. Many have wondered why Fahrenheit chose 32 degrees as the number for freezing and 212 for boiling.

Early on, Fahrenheit had been influenced by the Danish astronomer Olaus Roemer, who had established fixed points for the boiling point of water and the melting point of ice, which he labeled 60 degrees and 7.5 degrees, respectively. Fahrenheit based his own thermometer scale on Roemer's principles, but with some crucial changes. First, he did not want to work with "inconvenient and awkward fractions." Second, since Fahrenheit was primarily interested in measuring real atmospheric temperatures, he did not concern himself with the boiling point of water. A thermometer that would record temperatures that high was of no use in meteorology. Fahrenheit restricted his original measurements to what was actually observable in the atmosphere around him.

Fahrenheit set out to measure two fixed points: the temperature of water mixed with ice and sea salt; and the temperature of a healthy human body. The low point he labeled 0 degrees. The high point, the human body, he set at 96 degrees. Within this range he established water's freezing point at 32 degrees. Why these numbers? Fahrenheit originally used a 12-point scale and established 0, 4, and 12 as these three values. Then he refined his calibrations by introducing eight smaller gradations within each larger value. Thus, his temperature for the healthy human body, 96, was produced by multiplying 12 by 8. Later, when the boiling point of water was adopted as the standard high mark on the Fahrenheit scale, it was established at 212 degrees, or exactly 180 degrees above water's freezing point. This later required a retroactive adjustment of body temperature to 98.6 degrees.

TIME LINE

1686	1701	1714	1720s	1724	1736
Daniel Gabriel Fahrenheit born in Danzig (modern-day Gdansk, Poland).	Fahrenheit is apprenticed to a merchant to train as a bookkeeper.	Fahrenheit devises the first mercury thermometer.	Fahrenheit begins working with different kinds of thermometers.	Fahrenheit publishes the only account of his methods.	Fahrenheit dies.

LATER YEARS

Fahrenheit continued working in Amsterdam from 1717 until his death in 1736. In 1724 he published his only account of his methods in the *Philosophical Transactions of the Royal Society* in London, which elected him to membership that same year. In addition to the mercury thermometer and the temperature scale that bears his name, Daniel Fahrenheit first demonstrated that liquids other than water have fixed boiling points that are also affected by atmospheric pressure. He is also credited with developing a method of supercooling water—that is, cooling water below its normal freezing point without turning it into ice. Fahrenheit died on September 16, 1736, and was buried in The Hague, the Netherlands.

The Fahrenheit temperature scale continues to be used in the United States in the 21st century. Most of the world, however, uses the system developed by the Swedish scientist Anders Celsius. The Celsius scale, or centigrade system, is preferred by scientists, including those in the United States.

—Paul Schellinger

Further Reading

Book

Middleton, W. E. Knowles. *A History of the Thermometer and Its Use in Meteorology.* Baltimore, MD: Johns Hopkins University Press, 1966.

Web site

Temperature Converter
An online calculator for converting degrees Fahrenheit to Celsius and vice versa.
http://www.metric-conversions.org/temperature-conversion.htm

See also: Galilei, Galileo; Science, Technology, and Mathematics.

MICHAEL FARADAY

Inventor of the electricity generator
1791–1867

Some inventions develop by trial and error with little or no help from science. Others could never have come to light without painstaking scientific research; electricity is a good example. In the 18th century, electricity was a scientific curiosity; in the early 21st century, it powers most people's lives. This transformation is largely attributable to the invention of the electricity generator by Michael Faraday, one of the 19th century's greatest scientists.

EARLY YEARS

Michael Faraday, who died in a home owned by the queen of England, was born the son of a blacksmith on September 22, 1791, in what is now part of London. The family was extremely poor, not always having enough to eat. Faraday attended a day school where he learned the basics of reading, writing, and arithmetic; otherwise, he had little formal education. His family belonged to the Sandemanian church, a tight-knit community of Christians who believed, among other things, that God designed the natural world according to a very careful pattern. Religious beliefs like these greatly influenced Faraday's later scientific ideas and investigations.

At the age of 14, Faraday began working as an errand boy for a book-binder and soon learned his employer's trade. Suddenly exposed to hundreds of interesting books, he took the opportunity to educate himself. He loved reading scientific volumes, and after work he carried out his first experiments in chemistry and electricity. Self-betterment was the theme of Faraday's life at this time: one of his favorite books was Watts's *Improvements of the Mind.* In 1810, he joined the City Philosophical Society, a London club where young people met to discuss scientific topics and share knowledge—again with the aim of bettering themselves.

CHEMICAL PIONEER

In 1812, Faraday attended some public lectures at the Royal Institution, one of London's most prestigious scientific societies, where his favorite speaker was the distinguished chemist Humphry Davy (1778–1829). Faraday took notes during Davy's lectures, bound them attractively, and presented them to the great scientist, asking his hero whether he could come and work for him. Davy had no vacancies and warned Faraday against pursuing a poorly paid scientific career. However,

A portrait of Michael Faraday from 1860.

Faraday at work in his laboratory, located in the basement of London's Royal Institution. Painting by Harriet Moore, around 1852.

the following year, when a suitable opportunity arose, Davy remembered Faraday's enthusiasm and offered the young man a job as his assistant.

This was Faraday's big break, and he soon made himself indispensable. Beginning in 1813, when he was in his early twenties, Faraday accompanied Davy on a grand tour of Europe, where they met some of the world's greatest scientists, including Alessandro Volta (1745–1827). This great Italian physicist had explained how electricity could flow through cables as a current of power; he had also developed the world's first battery.

When they returned to England in 1815, Faraday continued to work as Davy's assistant at the Royal Institution. Faraday was soon making his own important chemical discoveries. In 1818, he helped to make harder forms of steel by combining iron with other metals to form alloys (mixtures of metals). Around 1820, he made the first known compounds of carbon and chlorine. Three years later, he became the first person to turn a gas into a liquid when he converted chlorine into liquid chlorine. In 1825, he discovered benzene, an important organic (carbon-based) compound. All this work earned Faraday a considerable scientific reputation. He was made a fellow of London's Royal Society in 1824 and became the director of the Royal Institution's laboratory in 1825. Years later, when Davy was asked to name his greatest scientific discovery, he replied, "Michael Faraday."

ELECTRICITY IN MOTION

Faraday's interest in electricity began in earnest when European scientists started making dramatic breakthroughs. In 1820, Danish physicist Hans

Christian Ørsted (1777–1851) found that when an electric current flows along a wire, it also makes a pattern of magnetism all around it. This was the beginning of electromagnetism—the theory that electricity and magnetism are two different aspects of the same underlying phenomenon.

The following year, French physicist André-Marie Ampère (1775–1836) made another breakthrough. He showed that if two parallel wires carry electric currents and are placed close together, the magnetism they make will push them apart or pull them together. This discovery—that electricity and magnetism can make force—led Faraday to his first great invention. In 1821, he took a battery, two jars of mercury (a metal that is a liquid at room temperature), a magnet, and a piece of wire. When he linked these elements into a circuit (a closed loop around which electricity flows), he found he could make a wire spin around a magnet or a magnet spin around a wire. His apparatus was a primitive forerunner of one of the most important technological advances of the 19th century: the electric motor (a device that uses electricity to turn a wheel, thus producing power to drive a machine or vehicle).

A diagram of Faraday's experiment, a forerunner of the electric motor. Electricity and magnetism are brought together to create force.

Faraday's Experiment Combining Electricity and Magnetism

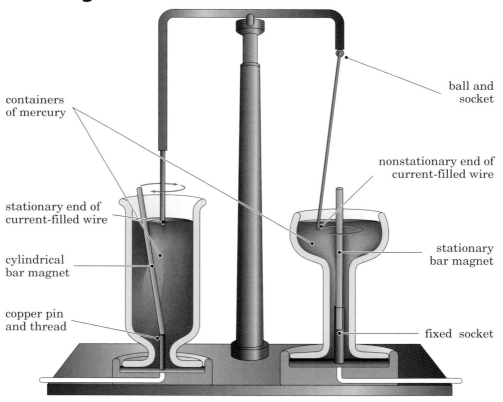

containers of mercury

ball and socket

nonstationary end of current-filled wire

stationary end of current-filled wire

cylindrical bar magnet

stationary bar magnet

copper pin and thread

fixed socket

MAKING ELECTRICITY

Exactly 10 years later, in 1831, Faraday made another important discovery about electricity—one that would have dramatic consequences. In an electric motor, electricity makes magnetism, and the magnetism produces motion. Faraday wondered if the opposite was true: was there some way of moving a magnet so that it would make electricity? Faraday thought so because he believed electricity and magnetism were closely linked. This was an example of how his religious belief in the harmony of nature influenced his scientific thinking: he thought God had made the world to an ordered pattern and a scientist's job was to reveal it.

To test his theory Faraday made a coil of wire and pushed a magnet in and out of it. To his delight, moving the magnet made electricity flow through the wire. He had discovered a way of making electricity simply by moving magnets and wires past one another. Thus he invented the electricity generator (or dynamo), a device that would produce a steady supply of electricity for as long as the magnets and wires were kept moving. In a related piece of work, Faraday also invented the electricity transformer. This consists of two separate coils of wire wrapped around an iron bar. If the coils contain different numbers of "turns" (wrappings) of wire, the device can increase or decrease the size of an electric current.

ELECTRICITY MEETS CHEMISTRY

A reproduction of the electrical dynamo built by French instrument maker Hippolyte Pixii in 1832. Pixii was the first to build a practical generator based on the principles demonstrated by Faraday.

Chemistry and electricity were the great themes of Faraday's scientific career, and he worked on them in tandem for much of his life. It was obvious to him that the two were connected and, following his invention of the generator, he set out to investigate how. In the 1830s, this led him to pioneer a field that became known as electrochemistry. His work included figuring out the laws of electrolysis, a way of splitting liquid chemicals by passing an electric current through them that Humphry Davy had pioneered.

During the 1830s, Faraday combined all his previous

research to come up with a new theory of electricity. He thought electricity was caused as "strains" built up in materials. When the strain became too great, an electric current flowed through the material in great waves, much as waves travel across the sea. Faraday's theory was a good contemporary attempt to explain what scientists knew about electricity, but his theory is now known to be false: electricity is really caused by a steady flow of tiny particles (electrons) that carry small charges (packets of electrical energy) through a material, somewhat like a line of ants carrying leaves.

LATER YEARS

By the end of the 1830s, as Faraday approached the age of 50, he was recognized as one of the most brilliant scientists in England. He had worked feverishly for almost three decades, pushing the boundaries of scientific knowledge. He had written several important books and many scientific papers. Like his mentor Humphry Davy, he gave dozens of lectures at the Royal Institution, inspiring many young people to take up scientific careers.

This astonishing output of work took its toll; in 1839, Faraday suffered a complete breakdown. With his health in ruins, for a time it seemed as though his career was finished. He began to devote himself to religion and became an elder of the Sandemanian church. Six years passed before he picked up his research again. However, in 1845, he returned to the world of science with his typical brilliance to make two more important discoveries in the area of electromagnetism.

One was the finding that magnetism can change the way light moves. If a beam of light is polarized (filtered so its waves move in only one direction), magnetism can make the light beam rotate (so its waves vibrate in a different direction). His other important finding was that crystals behave differently near magnetism. Some substances (which Faraday

The "great electromagnet" with which Faraday conducted many of his experiments is shown here, mounted underneath a bench in a laboratory.

America's Faraday

Michael Faraday's equivalent in the United States was Joseph Henry, born in Albany, New York, in 1797. Like Faraday, Henry came from a poor family and received little schooling. Whereas Faraday became an apprentice bookbinder, Henry became an apprentice watchmaker and silversmith. Like Faraday, Henry educated himself by reading. He entered Albany College at the age of 21; by the age of 28, he was professor of mathematics; just six years later, in 1832, he was a professor at Princeton, remaining there until 1846.

Also like Faraday, Henry was deeply curious about electricity and magnetism. It is now known that Henry discovered electromagnetic induction, the way magnetism can make electricity, about a year before Faraday. However, Henry was too busy teaching to publish his results, so Faraday gained the credit for the invention. In those days, scientists in different parts of the world were less aware of one another's work than they are today.

Henry was more of a practical inventor than Faraday. In 1829, he built the first working electric motor (Faraday's earlier motor was more of a laboratory experiment than a useful machine). Using his knowledge of electricity and magnetism, Henry also built some of the first electromagnets (powerful magnets operated by electricity, like those in a scrap yard). In 1831, he made the first electromagnetic telegraph and operated it over a distance of one mile—several years before Samuel Morse achieved the same thing. Henry is perhaps best remembered as the first secretary of the Smithsonian Institution. While there, he pioneered the use of the electric telegraph to send weather reports around the United States, an idea that led to the formation of the U.S. Weather Bureau in 1891.

Although Henry invented many things, he never patented any of them. He believed that science and technology should be used freely for the benefit of humankind. There are many striking parallels between the lives of Joseph Henry and Michael Faraday, and it is only an accident of history that makes Faraday the better known of the two today.

TIME LINE

1791	1815	1820	1823	1825	1831	1850	1867
Michael Faraday born in London.	Faraday works as Davy's assistant.	Faraday creates first compounds of carbon and chlorine.	Faraday is the first person to turn gas into a liquid.	Faraday discovers benzene.	Faraday invents electricity generator and transformer.	Faraday develops new theory of electro-magnetism.	Faraday dies.

called paramagnetic) are attracted to the magnetism and line up with it, whereas others are repelled by the magnetism and line up the opposite way (Faraday called them diamagnetic). In 1850, this work led Faraday to a whole new theory of electromagnetism as an invisible field that extends through space. He realized that the forces that act on objects in the world—gravity, electricity, and magnetism, for example—all work through this field. This was another example of how his religious belief in the unity of nature inspired his scientific ideas.

> Nothing is too wonderful to be true if it be consistent with the laws of nature.
>
> —Michael Faraday

Faraday's health began to decline in the 1850s and his mind was less sharp than earlier in his life. The Royal Society now refused to publish some of his experimental work, but the aging scientist still had his admirers. One of them was England's Queen Victoria. In 1858, she offered him the use of a house at her palace, Hampton Court, and wanted him to accept a knighthood also. Faraday graciously accepted the house, but declined the tremendous honor of a knighthood, modestly remaining Mr. Faraday until his death on August 25, 1867.

THE IMPORTANCE OF FARADAY'S WORK

Faraday's scientific research laid foundations for those who followed. His discoveries about electricity and magnetism helped another British scientist, James Clerk Maxwell (1831–1879), develop a complete theory of electromagnetism in the 1860s. Maxwell's scientific work and theory permitted the later development of radio, television, radar, cell phones, wireless Internet, and many other modern inventions. Faraday's theory of electric fields, which he called "ray vibrations," soon became known as field theory; today, it is regarded as one of the most fundamental ideas in modern physics.

Virtually all of Faraday's discoveries had a technological impact. The finding that electromagnetism can produce force led British inventor William Sturgeon (1783–1850) and American physicist Joseph Henry

(1797–1878; see box, America's Faraday) to develop practical electric motors. The opposite effect—force producing electromagnetism—enabled Faraday to invent the generator and the transformer. In the hands of Samuel Morse (1791–1872), Alexander Graham Bell (1847–1922), and Thomas Edison (1847–1931), these discoveries led society to the age of electric communication and power.

Faraday's chemical research was also technologically important. His work on strengthening steel was an early kind of metallurgy, the scientific study of metals for technological purposes. Metallurgy forms an essential part of many industries today, including aerospace and construction. Faraday's discovery of benzene was effectively the beginning of organic chemistry, a branch of science that led to the development of new artificial materials in the 20th century.

Michael Faraday's catalog of achievements is remarkable—all the more so because he was self-educated and from an underprivileged background. He owed his success to immense curiosity, hard work, utter determination, and unshakable religious beliefs. Luck played a part, as well: becoming an assistant to Humphry Davy set the stage for all that was to follow. Faraday's work demonstrates that the interweaving of science and technology can have an enormous impact on society.

—Chris Woodford

Undated painting by Alexander Blaikley. Faraday lectures at the Royal Institution in 1855; the two younger boys seated in the front row are the prince of Wales and the duke of Edinburgh.

Further Reading

Books

Archibald, Colin. *Michael Faraday: Physics and Faith*. New York: Oxford University, 2000.

Woodford, Chris. *Routes of Science: Electricity*. Farmington Hills, MI: Blackbirch, 2004.

Zannos, Susan. *Michael Faraday and the Discovery of Electromagnetism*. Hockessin, DE: Mitchell Lane, 2004.

Web site

Royal Institution of Great Britain

A detailed biography of Faraday, photographs of his apparatus, and transcripts of his lectures.

http://www.rigb.org/rimain/heritage/faradaypage.jsp

See also: Bell, Alexander Graham; Communications; Edison, Thomas; Energy and Power; Morse, Samuel; Science, Technology, and Mathematics; Volta, Alessandro.

PHILO FARNSWORTH

Inventor of electronic television

1906–1971

Some inventors dream of fame, others of fortune, others of changing the world. An inventor such as Philo T. Farnsworth, who developed a basic technology that could transmit pictures electronically, might well have achieved all three. Yet by the time Farnsworth died in the early 1970s, he was penniless and his contributions were largely forgotten. Only after his death was he recognized as one of the greatest American inventors of the 20th century and as the true "father of television."

EARLY YEARS

Philo T. Farnsworth was born into a family of Mormon farmers in a log cabin near Beaver, Utah, on August 19, 1906. Electricity, pioneered by Thomas Edison (1847–1931) in the 1880s, was just becoming widely available. Like many other young inventors, Farnsworth spent his early years tinkering with all kinds of farm machines and gadgets, many of them electrically powered.

When Farnsworth was 11, his family relocated to Rigby, Idaho, where he developed his first invention: a thief-proof lock. He later won a prize for this device in a national inventing contest sponsored by *Science and Invention* magazine. A voracious reader of science books and magazines, Farnsworth was captivated by the work of the brilliant German-born U.S. physicist Albert Einstein (1879–1955). By age 15, Farnsworth understood Einstein's photoelectric theory, which described the connection between light and electricity. Remarkably, Farnsworth could also explain Einstein's theory of relativity—one of the most complex scientific ideas ever proposed.

YOUNG INVENTOR

Farnsworth's breakthrough as an inventor came in 1922 as he guided a horse-drawn plow back and forth across his father's fields. As the machine plowed the soil into neat, parallel rows, Farnsworth realized that a picture—a pattern of light—could also be broken up into a series of parallel lines and transmitted by electricity. Greatly excited, the 15-year-old boy sketched the idea on a blackboard for his chemistry teacher, Justin Tolman. Although neither of them knew it at the time, this crude drawing would later play a vital part in the history of television.

Times grew harder for the Farnsworths; in 1922 they moved to Provo, Utah, a city

Philo Farnsworth in 1939 with some of the television equipment he invented (the woman in the photo is unidentified).

south of the state capital, Salt Lake City. Farnsworth began attending Brigham Young University, but only two years into his studies, his father died and Farnsworth had to leave to support the family. During this period, he took a variety of odd jobs, from working on the railroad as an electrician to cleaning the streets, but he continued his education by taking a correspondence course from the University of Utah.

How Television Works

Television is a form of communication that involves sending pictures from one place to another by electrical impulses. All objects reflect some of the light that falls onto them; that is why people can see them. In a broadcast studio, a television camera captures the light reflected off whatever is being filmed. The light is then turned into a sequence of electrical signals, using a scanning process similar to the one Philo Farnsworth originally devised.

With terrestrial television, the signals are converted into radio waves and beamed through the air using powerful antennas. Cable television sends these signals down fiber-optic cables; satellite television beams them down from orbiting spacecraft. Television sets in people's homes work in the opposite way to a television camera. First, they capture the incoming program signals from radio waves, fiber-optic cables, or satellites. Then they use the signals to build up, line by line, a pattern of light that matches exactly the pattern captured by the television camera, re-creating the image captured by the camera originally.

In 21st-century plasma televisions, tiny cells filled with gas are locked between two glass panels; the addition of electricity causes the gas to turn to a plasma, able to emit light.

Developing a machine to turn pictures into electricity remained Farnsworth's goal. His idea was to scan a beam of light across an image in parallel rows, like a horse plowing a field. The light would be reflected back into an electronic light detector (or "image dissector," as he called it) and turned into electrical pulses that could be transmitted down a cable. At the other end, a receiving apparatus could run a similar process in reverse, using the signals to power a scanning light beam that would draw a picture on a screen faster than the eye could perceive (see box, How Television Works). On January 7, 1927, he patented his "image dissector," which was effectively the world's first electronic television camera.

EXPERIMENTS WITH TELEVISION

While doing odd jobs for a charitable organization, Farnsworth met two professional fund-raisers, George Everson and Leslie Gorrell, who agreed to back his invention. The three men formed a partnership in 1926. Shortly afterward, Farnsworth married his girlfriend, Elma Gardner (nicknamed "Pem"), and they relocated to, where the development of electronic television began in earnest. A little over a year later, on September 7, 1927, and at only 21, Farnsworth made the first public demonstration of electronic television in a loft at 202 Green Street in San Francisco. In a scene reminiscent of the invention of the telephone, he stood in one room with the image dissector camera. Pem Farnsworth and George Everson were in another room with a primitive television receiver and watched as Farnsworth transmitted the first simple picture from one room to the other.

Development of the invention continued steadily during the late 1920s, with Farnsworth attracting growing media interest. By 1928, the San Francisco Chronicle was publishing reports of a "young genius" who was working on a "revolutionary light machine." A few years later, Collier's Weekly wrote about "electrically scanned television . . . destined to reach your home next year . . . largely given to the world by a . . . boy from Utah."

TIME LINE

1906	1917	1922	1927	1939	Late 1950s	1971
Philo T. Farnsworth born near Beaver, Utah.	Farnsworth wins a prize in a national inventing contest.	Farnsworth gets idea that eventually leads to invention of television.	Farnsworth demonstrates electronic television.	Farnsworth wins court battles and is awarded $1 million from RCA.	Farnsworth is one of the first people to research nuclear fusion.	Farnsworth dies.

WINNING THE BATTLE

Farnsworth was not the only person trying to develop television, however. In Britain, in 1925, a Scots engineer, John Logie Baird (1888–1946), had demonstrated an elaborate mechanical television system. Whereas Farnsworth used electronics for his image scanning, Baird opted for a clumsy, rotating wooden disc with holes cut into it that scanned an image mechanically. Farnsworth's goal was to "take all the moving parts out of television."

Lone inventors like Baird and Farnsworth were also attracting attention from large radio broadcasting companies, which began to fear that television, if successful, would harm their profits. Chief among them was the Radio Corporation of America (RCA), which dominated broadcasting in the 1920s. When its boss, David Sarnoff (1891–1971), discovered how close Farnsworth was to perfecting television, he feared the effect on RCA's business and decided to take action. He joined forces with a Russian-born television engineer, Vladimir Zworykin (1889–1982), who had been developing an alternative electronic television camera, the iconoscope, for the Westinghouse Company. In 1930, Zworykin visited Farnsworth's laboratory to see what his rival was doing; he did not tell Farnsworth that he was working for Sarnoff. Then Zworykin returned to his own laboratory and, with Sarnoff's financial backing, tried to improve on what he had seen at Farnsworth's lab.

When Zworykin was unsuccessful, Sarnoff changed his strategy: he approached Farnsworth and offered to buy him out for $100,000. Farnsworth had an idea how much his invention was worth and rejected the offer outright. Sarnoff was determined to defeat Farnsworth. He insisted that Zworykin was the rightful inventor of television because Zworykin had filed a patent for an iconoscope in 1923, four years before Farnsworth. However, because Zworykin had not demonstrated a working system, Farnsworth had gained the first patent. During the 1930s, RCA and Farnsworth fought a series of patent battles and appeals over their rival claims. Farnsworth was the ultimate victor in 1939; he was helped by the testimony of his high school teacher, Justin Tolman, who recalled the original blackboard sketch of the image dissector that his pupil had drawn for him in 1922. To Sarnoff's consternation, RCA was obliged to pay $1 million for a license to use Farnsworth's patents.

LOSING THE WAR

The battle with RCA had exhausted Farnsworth, leaving him depressed, in poor health, and an alcoholic. Yet that struggle was soon overshadowed by a far bigger conflict: the outbreak of World War II. As the war escalated, scientists and engineers throughout the United States became increasingly preoccupied with developing military technology, and interest in advancing television for entertainment declined.

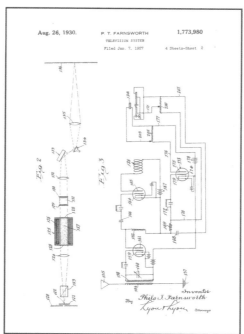
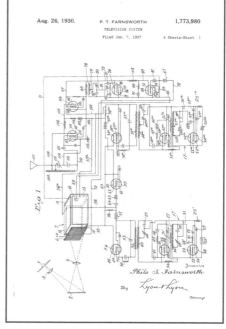

Illustrations from Farnsworth's 1927 patent application.

The Birth of Television

Television was developed first in Great Britain by Scotsman John Logie Baird. Baird made the first public demonstration of a television picture in a London department store in 1925. By 1928, Baird was broadcasting television pictures across the Atlantic, from London to New York City. The following year, the British Broadcasting Corporation (BBC) began regular television broadcasts using Baird's mechanical system.

In the United States, Baird's counterpart was inventor Charles Francis Jenkins (1867–1934), who had been trying to transmit pictures with electricity since 1894. Shortly after Baird's pioneering 1925 broadcast, Jenkins made the first public demonstration of television in the United States using a similar mechanical system. In June of that year, Jenkins was granted a U.S. patent for his system of "Transmitting Pictures over Wireless," and he set up the first U.S. television broadcasting station, W3XK, in 1928. Like the first home computers of the 1970s, Jenkins's early mechanical televisions were largely aimed at hobbyists, who built them from kits.

Farnsworth relocated to a farm in Brownfield, Maine, and concentrated on the Farnsworth Wood Products Company, which harvested timber for the war effort.

By the time the war ended in 1945, Farnsworth was out of touch with the electronics business and short of money. Rivals such as RCA, who had been manufacturing military equipment, began making consumer products once more. Farnsworth was still drinking heavily and was hospitalized for depression. In 1947, most of his patents expired just as television was finally becoming popular; RCA rapidly dominated the market, much as it had done with radio years before. In 1949, David Sarnoff hosted a special 25th-anniversary program to celebrate the birth of television. Disregarding Farnsworth's work, Sarnoff introduced Vladimir Zworykin as the technology's brilliant inventor.

LATER YEARS

The same year, now experiencing severe financial hardship, Farnsworth's company was bought out by International Telephone and Telegraph

(ITT). Philo Farnsworth remained at ITT for the next 18 years, where he was initially vice president of research and later a consultant. Although his health deteriorated, he developed a number of important military inventions, including an early warning system for missile defense and technologies that improved submarine detection and radar. Farnsworth also invented the first crude electron microscope and an incubator for newborn babies. Some of his inventions were more theoretical: in the late 1950s, he became one of the first people to research nuclear fusion (a way of generating energy by colliding small atoms so they join together), though his experiments were costly and ITT became increasingly reluctant to fund them.

In 1967, Farnsworth left ITT and moved to Brigham Young University, his old college in Utah. In recognition of his achievements, he was given an honorary doctorate of science and laboratory space, where he set up a new organization named Philo T. Farnsworth Associates (PTFA). Funding the expensive fusion research remained a problem, so members of the Farnsworth family sold all their stock, cashed in Farnsworth's life insurance policy, and borrowed from banks. It was not enough. By early 1970, the banks called in their loans and the IRS closed down the laboratory, seeking unpaid taxes. Penniless,

The landing on the Moon in 1969 was watched live on television by hundreds of millions of people around the world; it was also the event that convinced Farnsworth his work had been worthwhile.

U.S. Households with Television Sets, 1950 to 1975

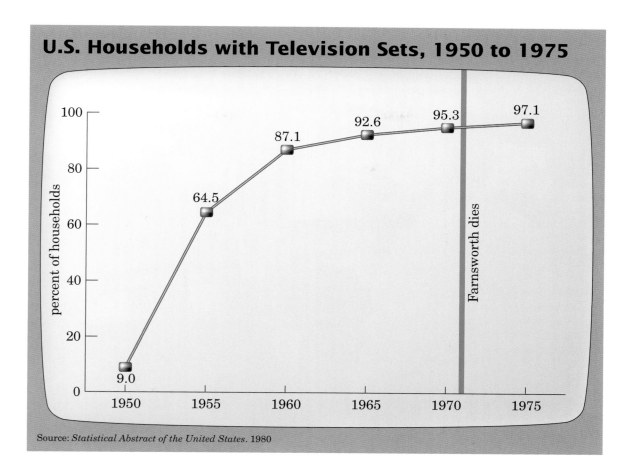

Source: *Statistical Abstract of the United States*. 1980

Unlike many inventors, Farnsworth lived to see his invention become a spectacular success.

depressed, and in poor health, Farnsworth contracted pneumonia shortly afterward and died on March 11, 1971.

By his death, Farnsworth held more than three hundred U.S. and foreign patents, making him a truly prolific inventor. Yet his most important invention, electronic television, had failed to bring him either recognition or fortune. Although he had hoped television would revolutionize human education, he loathed the way it became a cheapened form of entertainment, and eventually refused to let his children watch it. His wife recalled, "When Neil Armstrong landed on the Moon, Phil turned to me and said, 'Pem, this has made it all worthwhile.' Before then, he wasn't too sure."

In 1987, Pem Farnsworth unveiled a bronze statue of her husband in the Utah state capitol; it bears the inscription "Father of Television." The following decade, in 1999, *Time* magazine named Farnsworth one of the 100 most influential people of the 20th century.

—Chris Woodford

A customer inspects the many flat-screen televisions for sale in Niles, Illinois, in 2004.

Further Reading

Schatzkin, Paul. *The Boy Who Invented Television*. Terre Haute, IN: Tanglewood, 2004.

Schwartz, Evan. *The Last Lone Inventor: A Tale of Genius, Deceit, and the Birth of Television*. New York: HarperCollins, 2003.

Stashower, Donald. *The Boy Genius and the Mogul: The Untold Story of Television*. New York: Broadway, 2002.

See also: Communications; Entertainment; Young Inventors.

ENRICO FERMI

Inventor of the nuclear reactor

1901–1954

Everything in the world is made of minute particles called atoms. During the 20th century, scientists learned that breaking apart atoms can release energy. When this process happens again and again in what is known as a chain reaction, huge amounts of energy are produced in a short time. Chain reactions can be turned into practical technology using an apparatus called a nuclear reactor, invented in 1942 by Italian physicist Enrico Fermi. The first nuclear reactors were used to generate the massively destructive blast of the atom bombs in World War II; after the war, nuclear reactors were used to fuel nuclear power plants.

EARLY YEARS

Fermi was born in Rome, Italy, on September 29, 1901, and spent much of his life there. His father, Alberto, was a railroad official; his mother, Ida de Gattis, was a schoolteacher. Fermi had an older brother, Giulio; and an older sister, Maria. When Giulio died during a throat operation, Fermi, at the age of 14, was distraught and bought two old physics textbooks from a secondhand bookstall to cheer himself. He was so bright that he was soon correcting the mistakes he found in them.

At the age of 17, Fermi won a scholarship to the Reale Scuola Normale Superiore, a school linked with the University of Pisa; the essay he wrote for his entrance examination was judged to be outstanding and worthy of a degree in itself. Within a couple of years, he was teaching physics to his instructors. After he gained a doctorate from that university in 1922, studies outside Italy beckoned. Fermi spent several months studying with the distinguished physicist Max Born in Göttingen, Germany; then moved to Leiden in the Netherlands for a time; he then spent two more years in Italy, at the University of Florence.

In 1926, when he was only 25 years old, Fermi returned to Rome and was named professor of theoretical physics, a branch of physics that focuses more on math equations than on conducting experiments. By 1928, he had written *Introduction to Atomic Physics*, the first Italian textbook on modern physics, and met his wife, Laura Capon, a student at Rome University, where he was something of a legend. His brilliance was legendary, and his colleagues were soon calling him "the pope" because he seemed to be incapable of making a mistake.

MAKING AND BREAKING ATOMS

Atoms—and the extraordinarily complex world inside them—were Fermi's main focus for his entire life. In the 1920s, atomic physics, the science of how atoms behave, was still

Enrico Fermi, photographed at Columbia University in 1954.

TIME LINE

1901	1922	1926	1928	1934
Enrico Fermi born in Rome, Italy.	Fermi graduates from the University of Pisa with a doctorate in physics.	Fermi named a professor of theoretical physics.	Fermi writes *Introduction to Atomic Physics*.	Fermi discovers beta decay.

new. The field was born in 1896 when French physicist Antoine Henri Becquerel (1852–1908), accidentally discovered that the chemical element uranium gives off mysterious, invisible rays. This effect became known as radioactivity, and it suggested to scientists that atoms are really made up of smaller building blocks called subatomic particles. Radioactivity happens when unstable atoms give off particles or energy to make themselves more stable.

British physicist Ernest Rutherford (1871–1937) proved the hypothesis of subatomic particles in 1911 when he "split the atom": he fired atoms at one another until they literally smashed into bits. His results told him that atoms are largely made up of empty space, with most of their mass packed into a central area called the nucleus. In the years that followed, scientists found that the nucleus of an atom is built from two kinds of particles—protons and neutrons—and that other particles called electrons spin in the empty space around the nucleus.

The nucleus is really what makes an atom of one element, such as oxygen, different from an atom of another element, for example, carbon. If atoms were made from building-block pieces and could be broken apart, could they also be put together in new ways, and even changed into one another? Rutherford proved that they could: he changed nitrogen into oxygen by firing radioactive particles into nitrogen atoms. A French husband and wife, Frédéric (1900–1958) and Irène (1897–1956) Joliot-Curie, also changed elements into one another this way.

Science and technology are much like a relay race, in which discoverers and inventors build on work that has gone before and pass on their own ideas, in turn, to those who follow. Fermi picked up the baton from Rutherford and his colleagues and, from the 1920s onward, studied the way electrons behave. He soon produced an original mathematical theory of the world inside atoms; when a British physicist, Paul Dirac (1902–1984), arrived at similar results, their theories were combined and named Fermi-Dirac statistics. By 1934, Fermi had come up with a new way to explain one type of radioactivity, known as beta decay, which he said was caused when electrons escaped from unstable atoms.

TIME LINE (continued)

1938	1939	1942	1945	1954
Fermi wins Nobel Prize in Physics.	Fermi moves to New York and teaches at Columbia University.	Fermi's team produces the first nuclear chain reaction.	Manhattan Project team successfully tests the first nuclear weapon.	Fermi dies.

Next, Fermi built on the work of the Joliot-Curies: he tried to make some new radioactive elements by firing neutrons at atoms of different elements, including uranium. The idea was that the nucleus of one of these atoms would "capture" neutrons and turn into a different element.

One day, completely by accident, Fermi found that these experiments worked better if he did them on a wooden table rather than on a marble one. He was baffled before he realized that the atoms in the wooden table were slowing down the neutrons he fired—something that did not happen with the marble table. Slow neutrons were better than fast ones for making new elements , because a slow neutron spent more time near its target and was more likely to be captured. This work was very significant and Fermi won the Nobel Prize for Physics in 1938.

At first, Fermi and his colleagues believed he had made some new chemical elements in his neutron-firing experiments: they thought they had turned uranium into a series of bigger atoms by adding neutrons to its nucleus. Not until 1938, when three German scientists—Otto Hahn (1879–1968), Lise Meitner (1878–1968), and Fritz Strassman (1902–1980)—repeated the experiments was it clearly demonstrated that Fermi had not turned uranium into something bigger but actually split large uranium atoms into smaller bits. They called this process nuclear fission (another word for splitting). Meitner also realized that atoms are glued together by energy, so, when big atoms such as uranium break apart, enormous amounts of energy can be released. In the next few years, this remarkable discovery would change the world.

THE CHAIN REACTION AND THE BOMB

The world was already changing for quite different reasons. The growth of fascism (a violent political system that sprang up to oppose the spread of communism) in Germany and Italy led to the Holocaust, the persecution and slaughter of millions of Jews. That made Europe dangerous, especially for Fermi's wife, Laura, who was Jewish, and their two children, Nella (born in 1931) and Giulio (born in 1936). The chance to escape came in 1938, when Fermi was invited to Sweden to receive the Nobel Prize. Fermi and his family secretly planned to flee to the United States from Sweden rather than return to Italy.

By 1939, Fermi had a new job as professor of physics at Columbia University in New York City. There, Fermi repeated the experiment carried out by Hahn, Meitner, and Strassman. Discussing the result with a friend and colleague, the Danish physicist Niels Bohr (1885–1962), Fermi saw a new possibility. Suppose nuclear fission was designed so that each time an atom split, the pieces it broke into made other atoms split. These, in turn, would split more atoms. The result would be a runaway

The seventh level of the atomic pile, which Fermi tested successfully in December 1942.

The Nuclear Reactor

"The Italian navigator has landed in the New World."

On the chilly morning of December 2, 1942, a group of 43 of the world's leading scientists gathered around a strange piece of equipment that resembled a gigantic doorknob to carry out the world's most dangerous scientific experiment. Their leader, Enrico Fermi, had spent the last few weeks building this strange device inside a squash court under Stagg Field at the University of Chicago. Fermi called his apparatus an "atomic pile," and it was not hard to see why. It was, quite literally, a huge pile around 20 feet (6.1 m) high and 25 feet (7.6 m) wide, made up of about 45,000 very dirty black blocks of graphite (the material used in pencils). Holes had been drilled into the graphite and small cylinders of uranium put into them. Metal rods made of cadmium that could be raised or lowered into the pile by an electrical mechanism had been installed; more crudely, the emergency control rod could be released by chopping through a rope with an ax.

The atomic pile in Chicago was the world's first nuclear reactor and had just three parts. First, the uranium: this was the reactor's fuel and the source of all the energy it would produce. The graphite (known as a moderator) was designed to slow down nuclear reactions inside the pile to make them happen more effectively, just like Fermi's wooden table. The final part was the cadmium control rods. These were both the on-off switch and the accelerator pedal: they could make the nuclear reaction happen, make it faster, or turn it off completely.

When Fermi gave the order, someone flicked a switch and raised the cadmium control rods slightly in the pile. Now the unstable uranium atoms began to split, giving off neutrons. The neutrons passed through the graphite, which slowed them down, before they struck other uranium atoms and made them split.

An illustration (creator unknown) of the nuclear reactor test at a squash court at the University of Chicago in 1942.

Each fission (splitting) of a uranium atom caused more reactions. Soon the chain reaction avalanche was under way. Raising the control rods made the reaction faster; lowering them made it slow and eventually stop altogether.

Things proceeded smoothly on December 2, 1942, until a loud bang threw the scientists into a panic. Was the pile about to explode? One of the scientists noticed that the control rods had fallen into place and shut down the pile. Everything was fine. The scientists withdrew for lunch. By 2:00 p.m., they were back at the pile. Nuclear reactions were happening inside it, but there was no chain reaction. Fermi asked for the control rods to be lifted another foot to speed things up. Soon the nuclear reactions inside were going faster and faster. Around 3:23 p.m., Fermi made some calculations and announced: "The reaction is self-sustaining." He had proved that a chain reaction was possible. Twenty-eight minutes later, he ordered the control rods to be lowered and the pile to be shut down.

The 43 scientists celebrated by drinking cheap wine from paper cups. Shortly afterward, Arthur Compton, head of the U.S. nuclear weapons team, telephoned James Conant, chairman of the National Defense Research Committee, with the good news. They had to speak in code because of the immense secrecy of the Manhattan Project. Compton said, "The Italian navigator has landed in the New World." Fermi, in other words, had cracked the secrets of the nuclear chain reaction. However, the "New World" turned out to be a strange and dangerous place.

How the Atomic Pile Worked

Inside Fermi's atomic pile (nuclear reactor), many uranium atoms split up to release enormous amounts of energy.

A **fast-moving neutron** 1 passes through a block of **graphite** 2 (also known as a moderator), which slows it down.

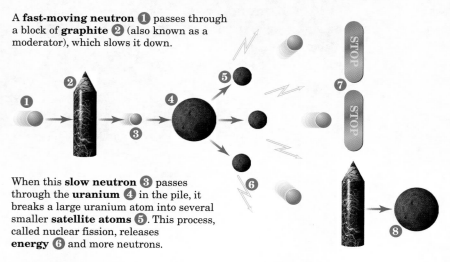

When this **slow neutron** 3 passes through the **uranium** 4 in the pile, it breaks a large uranium atom into several smaller **satellite atoms** 5. This process, called nuclear fission, releases **energy** 6 and more neutrons.

Some of the neutrons are soaked up by **control rods** 7 and stopped. The control rods can be raised or lowered in the pile to make the overall chain reaction happen faster or slower.

Other neutrons go on to split apart more **uranium atoms** 8 and release more energy.

reaction, a kind of atomic avalanche in which more and more atoms would break apart and vast amounts of energy would be released. Fermi realized that this chain reaction could form the heart of an incredibly powerful weapon: a nuclear bomb.

The late 1930s were tense times. The German dictator Adolf Hitler seemed unstoppable; troubles in Europe were about to plunge the world into six years of terrible conflict: World War II. A whole group of brilliant physicists and mathematicians, many of them Jews, had fled their European homelands for the safety of the United States. Concerned with Germany's growing military capacity, these scientists realized the dangers that lay ahead if Hitler made a nuclear bomb. On their behalf, one of their number, the famous physicist Albert Einstein (1879–1955), wrote to U.S. president Franklin D. Roosevelt (1882–1945) to persuade him that the United States must develop a nuclear weapon first.

Roosevelt agreed and immediately ordered a top-secret, $2 billion effort to produce an atomic weapon, code-named the Manhattan Project. Headed by J. Robert Oppenheimer (1904–1967), the project brought together many of the world's greatest physicists, including Fermi. His job was to show that a chain reaction really was possible. On December 7, 1941, the mission grew more urgent when the United States entered World War II following the Japanese bombing of Pearl Harbor in Oahu, Hawaii. The next year, Fermi moved to the University of Chicago. There, he made the world's first nuclear reactor, or atomic "pile" as he called it. Inside this strange piece of apparatus, on December 2, 1942, Fermi and his team produced the first nuclear chain reaction (see box, The Nuclear Reactor). One member of the team, Eugene Wigner (1902–1995), described the "eerie feeling" as it dawned on the scientists what they had done; soon after the experiment, another of the physicists, Leo Szilard (1898–1964), told Fermi it was "a black day in the history of mankind."

> Whatever Nature has in store for mankind, unpleasant as it may be, man must accept, for ignorance is never better than knowledge.
>
> —Enrico Fermi

During the next two and a half years, the U.S. government secretly constructed large-scale nuclear reactors at Oak Ridge, Tennessee, and Hanford, Washington, to produce nuclear materials for bombs. Elsewhere, others working on the Manhattan Project figured out how Fermi's chain reaction could produce a full-scale nuclear weapon. On July 16, 1945, at the Alamogordo Air Base in the New Mexico desert, the team successfully tested their first device. With the same power as 20,000 tons of high explosives, the weapon turned the sand in a two-mile circle around the bomb site into glass and sent a "mushroom" cloud seven and a half miles into the air.

Fermi was about nine miles from the test site and used a brilliantly simple method to calculate the bomb's power. He dropped strips of paper from his hand to the ground before the bomb was released. He did

the same thing after the impact, which—even at that distance—blew the papers slightly to one side as they fell. Measuring how far the papers had been knocked off course with a ruler, he calculated the bomb's energy—reaching almost exactly the same result as scientists who measured it with sensitive instruments. Weeks later, the United States dropped two nuclear bombs on the Japanese cities of Hiroshima and Nagasaki and World War II came to an end. "Little Boy," the bomb dropped on Hiroshima on August 6, 1945, flattened buildings within a 1.5-mile (2.4-km) radius of its impact and is believed to have led to around 200,000 deaths. "Fat Man," the bomb dropped on Nagasaki, killed approximately 70,000.

LATER YEARS

Enrico Fermi and his family had become American citizens in 1944. When the war ended, Fermi accepted a full-time professorship at the University of Chicago. There, he became director of the new Institute for Nuclear Studies and continued his research into atomic particles. Some

Fermi seated at the control panel of a particle accelerator in 1951.

of this work involved designing a huge new piece of apparatus called a synchrocyclotron, a powerful particle accelerator. This was a high-tech version of Rutherford's original atom-splitter: it sped up atomic particles, fired them at each other, and produced powerful collisions.

Fermi received many honors and awards for his work: a chemical element (number 100) was named fermium for him. The Enrico Fermi Award, given by the president of the United States, recognizes individuals who have made major contributions to the study or use of atomic energy. Fermi died in Chicago from stomach cancer, on November 28, 1954, at the age of 53.

THE WORLD AFTER THE BOMB

Few things have had such a dramatic impact on world politics as Fermi's discovery of the nuclear chain reaction. After the first experimental nuclear blast in 1945, the United States and other nations were soon building bigger and more destructive bombs. Within five years, the Soviet Union had built its own bomb. During the second half of the 20th century, nuclear weapons were created by other nations, leading to great concerns for world peace and security. Many people opposed their use, pointing to the way nuclear weapons had obliterated Hiroshima and Nagasaki with enormous loss of life. Others argued that nuclear weapons, though terrible, had saved more lives than they had taken, by bringing World War II to a quicker end. This debate continues today; some argue that nuclear weapons have prevented a third world war, and others are concerned about the harm that terrorists or rogue states could do with nuclear materials.

Nuclear reactors also have peaceful uses: they lie at the heart of nuclear power plants and also drive nuclear submarines and some spacecraft. During the 1950s, the United States, Canada, and Britain took the lead in developing nuclear energy. The first experimental plant, Calder Hall, opened in a remote part of northwest England in 1957; the United States opened a larger plant called Shippingport, near Pittsburgh, Pennsylvania, the same year. Although governments promised that nuclear energy would be virtually free, it turned out to be very costly. Nuclear plants proved expensive to build and hard to decommission (clean up) at the end of their useful lives. The disposal of radioactive waste and the risk of catastrophic accidents have also proved to be major public concerns, especially after serious accidents at the Three Mile Island plant in Harrisburg, Pennsylvania, in 1979 and at Chernobyl, Ukraine, in 1986. Even so, many countries still operate large nuclear programs: France, Belgium, and Sweden, for example, generate more than half their energy this way. The United States currently

The Fermi-2 nuclear reactor outside Detroit, Michigan.

has around 140 nuclear reactors; about one hundred are for power generation and the rest are for scientific research.

Today, the world has hundreds of nuclear reactors and thousands of nuclear bombs. All of them are based on the nuclear chain reaction that Enrico Fermi first demonstrated in 1942. Although Fermi played a central part in the Manhattan Project, he changed his mind about the more powerful nuclear-armed missiles (hydrogen bombs) made after the war, which he said were "weapons . . . of genocide." By then, he had no control over his discoveries or how people used them.

Some believe that scientists and inventors must always take responsibility for their discoveries, but most people who develop new technologies do not agree. They see their job as simply to push the boundaries of knowledge as far as they can, although they cannot always anticipate what effect their discoveries will have. New technologies often place heavy responsibilities on their creators because, once made, breakthroughs can never be "undiscovered." Fermi's work raises particularly strong passions because it led directly to the nuclear bomb—arguably the most destructive invention in human history.

—Chris Woodford

Further Reading

Books

Fermi, Laura. *Atoms in the Family: My Life with Enrico Fermi.* Chicago, IL: University of Chicago Press, 1995.

Richardson, Hazel. *How to Split the Atom.* New York: Franklin Watts, 2001.

Woodford, Chris, and Martin Clowes. *Atoms and Molecules: Routes of Science.* Farmington Hills, MI: Blackbirch, 2004.

Web sites

Kids' Zone: Nuclear Energy
   Atomic Energy of Canada's interactive guide to the energy locked inside atoms.
   http://www.aecl.ca/kidszone/atomicenergy/nuclear/puzzle.asp

The Manhattan Project: An Interactive History
   A U.S. Department of Energy project charting the creation of the atomic bomb.
   http://www.mbe.doe.gov/me70/manhattan/index.htm

Students' Corner: Nuclear Energy
   A guide by the U.S. Nuclear Regulatory Commission.
   http://www.nrc.gov/reading-rm/basic-ref/students.html

See also: Energy and Power; Military and Weaponry; Science, Technology, and Mathematics.

GEORGE FERRIS

Inventor of the Ferris wheel

1859–1896

Today's Ferris wheel is often dismissed as a quaint amusement park novelty, a mild diversion for those too old, too young, or too timid to join the thrill-seekers on roller coasters that reach heights of more than forty-five stories. However, at its debut at the 1893 world's fair in Chicago, the first Ferris wheel was considered a true marvel. For a time, its inventor, George W. Ferris, was one of the most celebrated engineers of his day.

EARLY YEARS

George Washington Gale Ferris was born in Galesburg, Illinois, on February 14, 1859. In early 1864, his family sold their dairy farm and cheese business and headed west by wagon. By that summer, however, the devaluation of the dollar during the U.S. Civil War had caused the Ferris family fortune to shrink by half. Concerned that their savings would disappear before they arrived in California, the Ferrises settled in Carson Valley, in the western part of the Nevada Territory.

The family stayed there for six years. It was a formative time for Ferris. Some believe he drew early inspiration for the Ferris wheel from a waterwheel at Cradlebaugh Bridge over the Carson River, imagining what it would be like to ride around and around in the wheel. By high school, he had moved to Oakland, California, to attend the California Military Academy. Ferris later attended Rensselaer Polytechnic Institute in Troy, New York, graduating in 1881 with a degree in civil engineering. His degree opened the door to a promising career designing bridges, tunnels, and train trestles throughout the Northeast and Midwest. He founded his own company, G.W.G. Ferris & Co., in Pittsburgh, Pennsylvania; the firm tested and inspected structural steel for railroads, mills, and bridges.

Ferris may have been inspired to design his wheel by watching waterwheels like this one.

G.W.G. Ferris & Co. soon expanded, opening offices in New York City and Chicago. Amid the success of his company, Ferris, could not have expected to become known, as one reporter put it, as "a wild-eyed man with wheels in his head."

TO OUT-EIFFEL EIFFEL

With the 1851 world's fair in London—the first of its kind—began the never-ending quest to make such fairs a showcase for the best, most impressive inventions in the world. The 1889 world's fair, in Paris, set a high bar; the most spectacular exhibit was the Eiffel Tower. The crowning achievement in the engineering career of Gustave Eiffel (1832–1923), the Eiffel Tower rose nearly one thousand feet into the sky.

> I remember remarking that I would build a wheel, a monster. . . . I fixed the size, determined the construction, the number of cars we would run, the number of people it would hold, what we would charge, the plan of stopping six times in the first revolution and loading, and ten making a complete turn. In short, before the dinner was over, I had sketched out almost the entire detail and my plan never varied an item from that day.
>
> —George Ferris

Daniel H. Burnham (1846–1912), director of the 1893 world's fair, to be held in Chicago, was determined to "out-Eiffel Eiffel." One Saturday in 1891, at an engineering club dinner, Burnham challenged the audience to construct something that would rival the Eiffel Tower, something "original, daring, and unique."

Ferris, then just 33 years old, rose to the challenge. According to legend, he drew up the specs for a "monster" observation wheel down to the tiniest detail, including ticket price, on gravy-stained napkins during dinner. The idea, presented the following day to the fair's building committee, fell flat. No one could fathom how such an enormous wheel might be constructed; nor did anyone trust that it could safely transport thousands of fair-goers. Fair officials dismissed Ferris as a "crackpot," but he was undeterred. He returned a few weeks later, having collected assurances from well-respected engineers that his design was feasible, as well as $400,000 for construction—$25,000 of his own money and the rest from local investors. Even so, time was needed to convince a wary building committee. When the committee finally gave Ferris approval, in December 1892, he had less than six months to bring his vision to reality; Eiffel, by comparison, had more than two years to build his tower.

THE MONSTER

Many have speculated about the source of Ferris's inspiration for his mammoth wheel. In addition to the popular theory about the Carson River waterwheel, writers and historians have pointed to bicycle wheels, merry-go-rounds, and European "pleasure wheels" or "ups-and-downs," which were small, crude, human-powered wheels dating back to the seventeenth century.

An "ups-and-downs" ride from seventeenth-century Turkey.

Ferris's design departed from each of these in scale. The Ferris wheel would measure 825 feet (251 m) in circumference and 250 feet (76 m) in diameter. It would rise nearly 265 feet (81 m) into the air. The wheel itself would be supported by two 140-foot (43-m) steel towers, each anchored in 30 feet (9 m) of concrete, designed to sustain five times the weight of the wheel and to withstand tornado-like winds of up to 150 miles per hour (241 km/hr). The wheel would turn on a 45-foot (13.7-m) axle—the largest single piece of forged steel in the world—and would be powered by two 1,000-horsepower coal-fired steam engines. It would have 36 luxurious wooden cars, each with its own uniformed conductor, to carry passengers through the air. In all, it would weigh 4,100 tons (3,719 metric tons).

The main pieces of the wheel were assembled at various steel mills in Detroit, then shipped on 150 railroad cars. They arrived in Chicago in March 1893, just two months before opening day of the world's fair on May 1. Construction had already been delayed because of problems with the foundation, including freezing winter temperatures and quicksand, and the project fell behind schedule. The wheel was not ready for opening day, but debuted six weeks later on June 21, 1893.

By all accounts, the Ferris wheel easily surpassed Burnham's hopes. In an effusive article of July 1, 1893, in *The Alleghenian*, one journalist wrote, "The Eiffel tower . . . when finished was a thing dead and lifeless. The wheel, on the other hand, has movement, grace, the indescribable charm possessed by a vast body in action."

The wheel could carry up to 2,160 passengers during each 20-minute ride—10 minutes spent loading the 36 gondolas, followed by 10 minutes of uninterrupted motion. On a clear day, passengers were said to be able to see three states from the top of the wheel. Few people of the late 19th century had been any higher than the rooftops of 12-story buildings, the "skyscrapers" of the day. The view from the Ferris wheel, at more than 25 stories, proved for many to be worth the 50-cent ticket, even though that price was a substantial portion of an average daily wage at the time.

Over the 19 weeks the Ferris wheel ran during the world's fair, it carried nearly 1.5 million passengers and grossed $726,805.50. At its peak, the Ferris Wheel carried more than 38,000 riders in one day. From most standpoints, the wheel was an overwhelming success—one that would change the face of amusement and recreation for generations.

A photograph of the Ferris wheel in Chicago in 1893.

THE DOWNWARD TURN

An illustration of the midway at the 1893 world's fair, featuring the first Ferris wheel in the background.

For George Ferris, the close of the world's fair marked the end of his celebrity; he dropped out of the spotlight almost as quickly as he had appeared. The more than $300,000 in profits from the wheel went not to Ferris but back to the board of the world's fair. The following spring, the wheel was moved, over the course of three months, to another part of Chicago at a cost of $14,000. In its new setting, the wheel lacked its initial charm and failed to bring in adequate revenue. Elsewhere, most notably in Coney Island, imitation wheels were being built, without payment or regard to the wheel's originator.

Ferris became increasingly obsessed with the demise of his wheel. He invested what was left of his savings in plans to build bigger and better wheels, but found no buyers. Convinced that the fair board had taken profits that were rightfully his, he spent the next two years in litigation. In early 1896, his wife left him. He moved into a hotel in Pittsburgh and died nearly bankrupt in Pittsburgh's Mercy Hospital on November 21, 1896—just 37 years old. The official cause of death is unclear, but has been listed variously as typhoid fever, tuberculosis, kidney disease, and suicide.

The death of the original Ferris wheel followed a decade later. It was sold at auction in Chicago, then shipped to St. Louis, Missouri, for the 1904 world's fair, where it earned just $250,000—not enough to cover the cost of shipping and assembly. On May 11, 1906, it was destroyed and sold for scrap. The remains, including much of the axle, are still believed to be buried beneath the original fairgrounds, in what is now Forest Park in St. Louis.

COMING FULL CIRCLE

By the end of the 19th century, other wheels had been constructed in Vienna and London. In 1906 William Sullivan, an engineer in Illinois, formed the Eli Bridge Company to produce the smaller-scale—and, more important, portable—wheels still in use at local fairs today.

More recently, fairgrounds and amusement parks have seen a resurgence of wheels that pay homage, in scale and magnitude, to the work of George Ferris. In the United States the Texas Star reaches 212 feet (65 m). Japan boasts several giant wheels: the Tempozan Harbor Ferris wheel and the Sky Dream Fukuoka, both nearly four hundred feet, and the Cosmoclock, at 344 feet (105 m) tall. The London Eye rises 443 feet (135 m) above the Thames River. Plans for wheels reaching 500 feet (152 m) or more are under way in Singapore, Shanghai, Las Vegas, and Moscow. More than one hundred years after it debuted in Chicago, the Ferris wheel has reached new heights, both literally and figuratively.

—Laura Lambert

The London Eye, located on the South Bank of the Thames River in London, England.

TIME LINE

1859	1881	1891	1893	1894	1896
George Ferris born in Galesburg, Illinois.	Ferris founds his own company, G.W.G. Ferris & Co., in Pittsburgh, Pennsylvania.	Ferris accepts the challenge to construct something for the 1893 world's fair.	Ferris's wheel debuts successfully, six weeks after the world fair's opening day.	Ferris embarks on a series of lawsuits against the world's fair board.	Ferris dies in Pittsburgh, Pennsylvania.

Further Reading

Books

Anderson, Norman D., and Walter R. Brown. *Ferris Wheels*. New York: Pantheon, 1983.
Jones, Lois Stodiek. *The Ferris Wheel*. Reno, NV: Grace Dangberg Foundation, 1984.

Web site

Carnegie Library of Pittsburgh: George Ferris
Features two archived newspaper stories about Ferris from 1893.
http://www.clpgh.org/exhibit/neighborhoods/northside/nor_n105a.html

See also: Entertainment.

FOOD AND AGRICULTURE

Earth's population has increased at a dramatic rate. During the first millennium (ending in the year 1000 CE), population rose by 75 million, but in the second millennium (ending in 2000) the increase was six billion—around 80 times greater. A planet with finite resources can support only so many human beings, but a long series of agricultural inventions have so far enabled the world to feed the growing population. Whether technology can continue providing for the world as the world's population continues to rise remains to be seen.

EARLY HISTORY

Agriculture, which means cultivating land to make food and other products, began in the Middle East approximately ten thousand to eleven thousand years ago. Before this time, most people lived as hunter-gatherers, getting their food whenever they needed to, much as animals do, by hunting and fishing or gathering wild plants. Agriculture offered a more systematic way of producing the foodstuffs people needed to survive. Societies gradually began to domesticate animals, plant crops,

Some of the earliest farms were terraced into hillsides, and in the 21st century some farms are still built this way, as in the Italian farm shown here.

and develop better tools, includings picks; digging sticks for sowing seeds, cultivation, and harvest; and plows, specifically to help them farm the land. Many basic farming techniques developed in this time are still used. For example, hillsides were first terraced (cut into steps to make level growing areas) in Peru and other countries thousands of years ago, whereas farmers were making attempts to select plants that could resist diseases about nine thousand years ago.

Growing plants need huge amounts of water so, in arid areas like the Middle East, moving water from lowland rivers to crops on higher ground was a major endeavor. Some of the earliest middle eastern inventions are devices for raising and moving water, developed in ancient times. Around 1400 BCE, the ancient Egyptians invented a type of water crane known as a shadoof. Working in a manner similar to a seesaw, it had a bucket at one

end and a heavy weight at the other. Using a series of shadoofs, Egyptians could raise water through all the levels of a terrace. A more efficient water-raising device was invented by ancient Greek engineer and mathematician Archimedes (ca. 287–212 BCE). The device was a giant screw that drew water up a hollow pipe when the operator cranked a handle. The Roman Empire (27 BCE–395 CE) introduced middle eastern farming techniques to Europe and pioneered irrigation canals and aqueducts for carrying water to farms.

THE MIDDLE AGES

Some of the simplest agricultural inventions have had the greatest impact. When horses were used as work animals in ancient times, loads were attached to straps around their necks. During the Middle Ages (the period from the fall of

An undated illustration of Jethro Tull's seed driller.

the Roman Empire to about the fifteenth century), European farmers began using harnesses that distributed the load more evenly around the animal's body. This simple improvement enabled the horse to breathe more easily and work harder, so it could pull a plow several times faster than an ox. Such harnesses greatly increased productivity and helped to make individual farmers more wealthy and independent.

Toward the end of the Middle Ages, explorers set off on great ocean voyages to conquer the world. In the Americas, explorers were surprised to find Native Americans already using advanced agricultural techniques, and the Europeans took the foods they discovered back to their homelands. In this way, European countries learned of foods and other goods such as chocolate, corn (maize), potatoes, rubber, tobacco, and tomatoes from the Americas, and coffee and tea from Asia.

MECHANIZING FARMS

A new phase in the history of agriculture began in Britain around 1700 and spread throughout Europe and North America during the next two centuries. Better ways of growing crops and improved methods of breeding livestock were two of the forces behind the agricultural revolution, as this period became known. The third major force was an improvement in farm machinery. One of the pioneers of mechanized farming was English farmer Jethro Tull (1674–1741), who developed an improved drill for pushing seeds into the

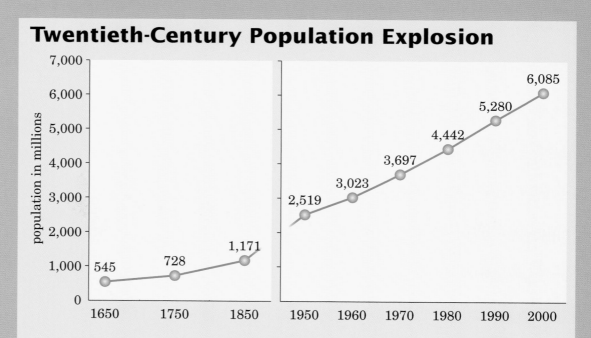

Twentieth-Century Population Explosion

population in millions

545 — 1650
728 — 1750
1,171 — 1850

2,519 — 1950
3,023 — 1960
3,697 — 1970
4,442 — 1980
5,280 — 1990
6,085 — 2000

Source: (1650–1850) A. M. Carr-Saunders, 1936. *World Population*, Norwood Editions, Oxford. (1950–2000) Population Division of the Department of Economic and Social Affairs of the United Nations Secretariat, *World Population Prospects: The 2004 Revision* and *World Urbanization Prospects: The 2003 Revision*, http://esa.un.org/unpp, accessed October 11, 2006.

Modern agricultural methods have led to an increase in the amount and the quality of available food, which was one important factor in the explosive population growth in the 20th century.

ground in 1701. Tull's invention made the sowing of seed more efficient, thus reducing the amount of seed needed to achieve the same acreage of crop. He also developed a hoe, drawn by horses, that ripped up weeds and aerated the soil as it was pulled through the fields.

Tull had not invented the idea of drilling seeds—which dates from ancient times—but he used technology to do the same job more quickly and efficiently. John Deere (1804–1886), a blacksmith from Illinois, was another inventor who refined existing technology. In 1837 he invented a new form of plow made from highly polished steel: a toughened form of iron popularized by Henry Bessemer (1813–1898). It was more effective at moving through heavy soil than earlier plows made from iron and wood, and played an important role in the expansion of agriculture in the Midwest.

Machines were also invented for the harvesting of crops. One of the pioneers was Cyrus Hall McCormick (1809–1884), who developed a horse-drawn reaping machine in 1831 for cutting grain. Later that decade, the brothers John and Hiram Pitts of Winthrop, Maine, patented the first thresher (a machine that separates the useful part of a crop, such as the grain; from the unwanted part, which is known as chaff). Combine harvesters were also invented in the 19th century. As their name suggests, they worked by combining reaping and threshing in a single machine. However, they were not widely used for many decades.

A man sits on a Fordson Model F tractor in a photograph from around 1920.

A man sprays pesticides on grape leaves at a California vineyard in 1943, before the dangers of pesticide exposure were fully realized.

POWERING THE REVOLUTION

All these early machines were horse-powered, but several new ways of driving farm machines emerged after the mid-19th century. Steam power was pioneered by English blacksmith Thomas Newcomen (1663–1729), who built his first, huge steam engine for pumping wastewater out of a deep coal mine in 1712. Smaller steam engines were later used in railroad locomotives and ships. By the mid-19th century, inventors were using stationary steam engines on farms, mainly to drive threshing machines. Traction engines (large tractors powered by steam) became popular in the decades that followed.

Steam engines were cumbersome and inefficient, but better power sources soon appeared. German inventor Nicolaus August Otto (1832–1891) invented the

gasoline engine in 1867, and his compatriot Rudolf Diesel (1858–1913) developed his doubly efficient diesel engine in the 1890s. Gas and diesel engines, also known as internal combustion engines, played an important part in the continuing agricultural revolution as they came to be used in cars, trucks, and tractors. Automobile pioneer Henry Ford (1863–1947), who was raised on a farm, helped many more farmers to afford tractors when he introduced his inexpensive Fordson Model F in 1917. Aircraft also played a part in modern agriculture, aiding activities such as crop dusting. They were developed after American brothers Orville (1871–1948) and Wilbur Wright (1867–1912) mounted a gasoline engine on a glider to make the first powered flight in 1903.

During the 20th century, electricity became increasingly important in powering farm machines. Electric power became available during the 1880s, when the American inventor Thomas Edison (1847–1931) built his first power plant in

According to the U.S. Department of Agriculture, Americans spent just over $9 billion of their disposable income on food in 2005. While that number may sound large, it represents only 9.9 percent of Americans' total disposable income for that year. As disposable income climbs, the percentage spent on food drops: in 1970, Americans spent 13.9 percent of their disposable income on food; in 1950, 20.6 percent.

New York City. Most farms in Europe and the United States were using electric power by the end of the 1930s.

IMPROVING LAND AND CROPS

No amount of machinery can force crops from poor land, so the development of scientific methods of farming in the 19th century was just as important as the arrival of engines, tractors, and electricity. A pioneer in this field was the English chemist Humphry Davy (1778–1829), who explained his ideas on using chemicals to fertilize plants in 1813. Three decades later, in 1842, English scientist John Bennet Lawes (1814–1900) patented a way of fertilizing crops with phosphates, the basis for the huge industry that grew up to develop and provide artificial fertilizers.

Another form of treating land chemically, using pesticides to kill insects, became possible with the discovery of dichlorodiphenyltrichloroethane (DDT),

a chemical originally developed to kill mosquitoes. In 1942, Swiss chemist Paul Müller (1899–1965) found that DDT was an effective way of killing a wide variety of agricultural pests, such as the louse and the Colorado beetle. A whole series of similar pesticides were developed from DDT in the years that followed. Initially, such chemicals seemed to offer a cure-all: they killed pests, prevented the spread of diseases such as typhoid and malaria, and saved crops and lives. However, in the 1950s and 1960s, these chemicals were found to be immensely toxic and very persistent in the environment and in the bodies of animals and humans. Some, including DDT, were later banned.

Botany (the study of plants) began in ancient times when the Romans were the first to breed plants selectively by carefully picking the best examples and cultivating them. Only in the 19th century, however, did selective plant breeding begin to develop into a truly systematic science. Agricultural scientists at this time raised

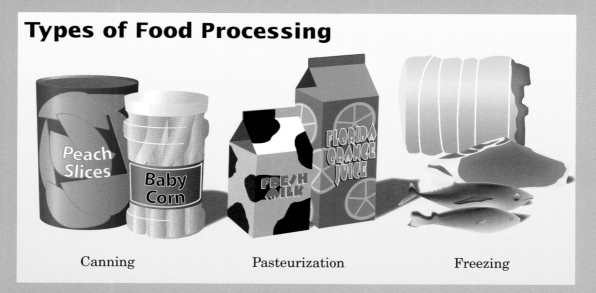

Types of Food Processing

Canning   Pasteurization   Freezing

Much of the food consumed in contemporary western nations undergoes some form of processing. Canning, pasteurization, and freezing are the three major types.

Fear of the Future

Making enough food for an increasing population has often involved controversial practices, although people have not always seen them as such. Slave labor, for example, was normal in ancient Rome, where slaves ran the huge farms that fed the Roman Empire and even powered many of its machines by walking endlessly on or inside treadmills. Slavery was not widely considered unacceptable until the 19th century, when Africans were being imported into the southern United States in the tens of thousands to enable the growth of the cotton industry.

Sometimes the controversial aspects of new technologies are not immediately apparent. Chemical pesticides (insect killers) and herbicides (weed killers) seemed to offer farmers a miraculous way of increasing crop yields when they first appeared during World War II. Then, in 1962, ecologist Rachel Carson (1907–1964) published a damning book, *Silent Spring*, that documented what she saw as the dangers to plants, animals, and humans. Passionately opposed to the new technology, she argued: "Chemicals are the sinister and little-recognized partners of radiation in changing the very nature of the world—the very nature of its life."

Developments in biotechnology since the 1970s have caused a whole new series of controversies. Some scientists believe genetically modified (also known as GM or transgenic) organisms offer the best hope of feeding the earth's growing population. Yet many environmentalists fear that GM technology could contaminate natural plants, produce "super-pests" that cannot be controlled, and have other unforeseen consequences. In the United Kingdom, protesters have repeatedly uprooted GM test crops; in India, farmers have burned GM seeds and marched in protest.

Sometimes fears of the future are justified. In 1800, around 90 percent of the U.S. population worked in agriculture, whereas now that figure is less than 2 percent—so it is certainly true that technology has eliminated millions of agricultural jobs. Some would see this as the price of progress. The question is not whether technology is good or bad, but whether the benefits of technology outweigh its costs—to people's lives and to the environment.

At the International Rice Research Institute in the Philippines, biotechnologist Swapan Datta tends to genetically modified plants in 2003. So-called golden rice has been genetically modified to increase its beta-carotene content.

contributed to saving hundreds of millions of people in developing countries from starvation and earned him the Nobel Prize for Peace in 1970. Such advances became known as the "green revolution," the idea that agricultural innovations could solve the problem of world hunger.

plants in laboratories and carried out experiments to develop better varieties (slightly different versions of the same plant species). Luther Burbank (1849–1926), for example, developed improved varieties of many plants, including potatoes, plums, roses, corn, and tomatoes, and in 1921 set out his ideas in a book called *How Plants Are Trained to Work for Man*. Burbank's title could have been a personal motto for the life of George Washington Carver (1864–1943), an African American inventor who developed hundreds of useful products from peanuts and sweet potatoes.

Developing improved varieties of crops has also played a crucial role in the battle to feed the world. In the mid-20th century, U.S. agricultural scientist Norman Borlaug (1914–) developed new varieties of wheat and other grains that were more resistant to disease and produced far higher yields. Borlaug's work

PRESERVING AND CONSUMING FOOD

Produce often endures long periods between leaving the farm and arriving in grocery stores or homes. As transporting food grown in distant places has become easier, preserving food has become increasingly crucial. Most of the foods people eat in western countries are preserved in some way.

A French chef, Nicolas-François Appert (ca.1750–1841), was one of the first inventors to tackle food preservation. In 1810, he invented a way of heating and canning food to keep it safe to eat for longer periods. This technique made possible the development of industrial canning during the 19th century. Inventors such as Appert knew how to preserve food, but not why it spoiled or how their methods worked. Answers to those questions also came from France. Louis

TIME LINE

ca. 9000-8000 BCE	ca. 1400 BCE	1701	1712	1800s	1810
Agriculture begins in the Middle East.	Egyptians invent the shadoof.	Jethro Tull develops a drill to improve land productivity.	Thomas Newcomen builds the first steam engine to pump wastewater from coal mines.	Agricultural scientists begin to study selective plant breeding.	Nicolas-François Appert invents a method of heating and canning food.

The organic section of a supermarket in Des Plaines, Illinois, in 2006. Consumers' interest in organic foods developed so quickly in the early 21st century that the demand often outpaced the supply.

TIME LINE

1813	1831	1837	1842	1880s	1894
Humphry Davy explains how to use chemicals to fertilize plants.	Cyrus Hall McCormick develops a horse-drawn reaping machine.	John Deere invents a new type of plow out of steel.	John Bennet Lawes patents a way to fertilize crops with phosphates.	John Pemberton invents Coca-Cola.	Will K. Kellogg invents cornflakes.

Pasteur (1822–1895), a French chemist and biologist, discovered that food spoils because of bacteria growing inside it. He developed a method of heating food (pasteurization) that kills bacteria and makes food safe longer. An alternative means to the same end is to cool food so that the bacteria grow more slowly. In the 1920s, American inventor Clarence Birdseye (1886–1956) perfected a way of quickly freezing foods such as fish so they could be kept fresh longer. Canning, pasteurization, and freezing are still the most important methods of preserving food, along with newer and more controversial technologies such as irradiation (using radioactivity to kill pests and bacteria).

Inventors have found many other ways of satisfying consumers' demand for delicious, different, and out-of-season foods. Modern breakfast cereals were invented in 1894 when Will K. Kellogg (1860–1951) and his brother John forced boiled wheat through rollers and accidentally discovered cornflakes. The use of Coca-Cola, invented in the 1880s by John Pemberton (1831–1888), as a soft drink was also accidental. Originally designed as a medicinal drink, Coca-Cola was soon being bought solely because people liked its taste. Another important way to preserve foods is simply to make it easier and more convenient for people to shop for them. Clarence Saunders (1881–1953) devoted his life to developing self-service stores. His Piggly Wiggly chain, founded in 1916, effectively launched the idea of the modern supermarket. Such stores have had a dramatic impact on farming. By buying in large quantities, chain stores have made importing food from all over the world economically feasible. Whereas grocery stores would once have stocked only foods grown during a particular season, now they can sell virtually any food all year round.

MODERN FARMING

Science and technology have brought great advances in agriculture and food production. Engineering has developed better engines and machines, agricultural science has given farmers more productive fields, and chemistry has helped to conquer pests and weeds. All these trends will continue, but biotechnology is likely to be the most important source of new breakthroughs. Biotechnology broadly means using human ingenuity to harness natural, biological processes—encompassing everything from brewing and making bread (age-old examples of biotechnology) to

TIME LINE (continued)				
1916	**1917**	**1920s**	**1921**	**1930s**
Clarence Saunders founds the Piggly Wiggly store chain.	Henry Ford introduces the Fordson Model F tractor.	Clarence Birdseye perfects method of flash-freezing foods.	Luther Burbank writes *How Plants Are Trained to Work for Man*.	Most European and American farms begin to use electric power.

Activists in 2003 protest the growing of genetically engineered corn in Villanueva de Gállego, Mexico. The slogan on the banner reads, "Stop genetic pollution."

cloning animals and genetically modifying crops (the latest examples).

Cloning, invented in the early 1970s by American scientists Herbert Boyer (1936–) and Stanley Cohen (1935–), involves manipulating DNA (deoxyribonucleic acid, the basic genetic instructions that tell cells how to grow) to make identical copies of an organism. Originally used to manufacture drugs like insulin (a treatment for diabetes), cloning has recently been applied to make genetically identical "copies" of farm animals. Dolly the sheep, the world's first cloned animal, was born in Scotland in 1996. Named for

the singer Dolly Parton, she was created by a team of biologists led by Englishmen Keith Campbell (1954–) and Ian Wilmut (1944–).

A closely related technology, genetic modification, first came into widespread use in the 1990s. It involves altering the DNA of a plant or animal to give the organism some added benefit. For example, one of the first genetically modified (GM) products, a tomato dubbed "Flav Savr," introduced in 1994, was engineered to ripen more slowly and last longer in stores. Techniques such as GM foods and cloning have proved intensely controversial (see

TIME LINE

1942	Mid-1900s	Early 1970s	1996	1994
Paul Müller finds DDT can be used to kill agricultural pests.	Norman Borlaug develops new disease-resistant and higher-yield plant varieties.	Herbert Boyer and Stanley Cohen develop the cloning process.	Keith Campbell and Ian Wilmut lead the team that produces Dolly the cloned sheep.	The Flav Savr tomato is launched.

box, Fear of the Future). Yet, if the ethical issues can be settled, they may offer a way of delivering more and better farm products for earth's ever-growing population—and a way of continuing the agricultural revolution long into the future.

—Chris Woodford

Further Reading

Books

Goldberg, Jake. *The Disappearing American Farm*. New York: Franklin Watts, 1996.

Hadley, Ned. *Eyewitness: Farm*. New York: Dorling Kindersley, 2000.

Miller, Char, ed. *Atlas of US and Canadian Environmental History*. New York: Routledge, 2004.

Wilkes, Angela, and Eric Thomas. *A Farm Through Time*. New York: Dorling Kindersley, 2002.

Web sites

Harvest of Fear
An introduction to the modern biotechnology debate from PBS.
http://www.pbs.org/wgbh/harvest/

USDA
Information and activities from the U.S. Department of Agriculture.
http://www.usda.gov

See also: Appert, Nicolas-François; Birdseye, Clarence; Borlaug, Norman; Burbank, Luther; Carver, George Washington; Deere, John; Gadgil, Ashok; Kellogg, Will Keith; McCormick, Cyrus Hall; Pasteur, Louis; Pemberton, John; Saunders, Clarence.

HENRY FORD

Inventor of the Model T and
moving assembly line

1863–1947

Automobiles changed society dramatically during the 20th century, and
Henry Ford played a major part in that transformation. Although Ford did
not invent cars, he developed a way of producing them less expensively, so
more people could afford them. In broader terms, Ford's style of business
helped to transform the United States into an overwhelmingly industrial
nation, gave people more opportunities to own manufactured goods, and
greatly increased their wealth and leisure time.

EARLY YEARS

Henry Ford's childhood was typical of a boy living in the United States in the mid-19th century. He had been born in Greenfield, Michigan, on July 30, 1863, and he spent his days in a one-room schoolhouse and his evenings doing chores on the family farm. Machinery fascinated him; his mother called him a "born mechanic." At age 12, he saw his first self-propelled vehicle, a giant coal-powered tractor, which seemed to him to be a messenger from the future. Ford would never forget the experience, later saying, "It was that engine which took me into automotive transportation."

Ford's family had been farmers for generations and it was always assumed that he would take over the farm, but he had other ideas. At the age of 16, he walked to Detroit and arranged an apprenticeship with a mechanic. He also worked evenings in a watch-repair shop. He

Henry Ford photographed in 1888.

considered becoming a watch manufacturer until he realized that watches were a luxury, not a necessity. "Even then," he later remembered, "I wanted to make something in quantity."

For the next few years, Ford divided his time between engineering jobs and the farm. On the farm, he erected a machine shop and built a small, experimental "farm locomotive" to mechanize the chores he hated so much. In 1884, his father promised him some timberland if he gave up being a mechanic. After meeting Clara Jane Bryant, he grudgingly agreed to his father's request so they could afford to wed. Ford married Clara in 1888 and their only son, Edsel, was born five years later.

ENGINEER

Marriage brought responsibility, and it induced Ford to take a steadier job. In 1888, he became an engineer with the Detroit branch of the Edison Illuminating Company, the firm Thomas Edison had established to bring electric light to the nation. Ford's job was to keep the power on, but he often had little to do. It was the perfect job for an inventor, and he spent the next five years trying to develop a practical gasoline-powered engine. After he succeeded, he was introduced to Thomas Edison at a company gathering; Edison praised Ford and encouraged him to "keep at it."

Having developed an engine, Ford next wanted to build an automobile. This he managed in 1896. The car he developed was not at all like modern cars. Named the Quadricycle, it ran on four 28-inch bicycle wheels with rubber tires and was steered by a large moving tiller at the front. When Ford tried out this "gasoline buggy" on the streets of Detroit, crowds gathered and terrified horses bolted in all directions.

Ford takes a ride on his Quadricycle in 1896.

MANUFACTURER

Ford realized that if he could make one car, he could make many, but his hobby conflicted with his job at Edison. The company gave him a choice: it would promote him to general superintendent or he could go to work for himself. Confident, and with his wife's support, Ford chose the latter. In 1899, supported by a few wealthy backers, he formed the Detroit Automobile Company. For three years, Ford built luxury cars to order, but he wanted to make less expensive models that most people could afford.

In 1902, Ford decided to go his own way, and the following year he launched a new firm—the Ford Motor Company. By quitting his previous firm, he angered his investors—the very people who might have supported the new company. As a result, Ford had only $28,000 in capital, most of it provided by ordinary citizens. In its first year of operation, the Ford Motor Company made a simple vehicle called the Model A, costing $850, and sold 1,708 of them. In 1908, after selling thousands of various models, Ford introduced his first truly affordable vehicle: the Model T. From then on, only one simple model would be offered and as Ford famously said: "Any customer can have a car painted any color that he wants so long as it is black." The customers loved it, and Ford sold cars by the million (see box, The Model T Ford).

INDUSTRIALIST

Ford's secret of success was mass production. His cars were gradually assembled by passing down a whole line of workers, a configuration known as an assembly line. The idea for the assembly line dates back to the 19th century. Eli Whitney (1765–1825) had developed a way of manufacturing rifles from standard pieces, a method that became known as mass production. Unlike the practice of a craft, in which one skilled worker makes one product from start to finish, mass production involves a group of workers, each of whom makes or handles only one piece of the product.

In 1913, Henry Ford took mass production to a new level by creating a conveyor belt pulled by a chain along the factory floor. The automated mechanism carried semi-completed car bodies past the line of workers who were putting them together. Each worker on the assembly line had a very specific job to do and an exact amount of time in which to do it if the whole system was to run smoothly.

Ford realized that automation was the key to manufacturing, but he also knew people were crucial to automation. Mass production was more tedious for workers than traditional craft work. Assembly workers

Ford's first moving assembly line, installed in the Highland Park factory in 1913.

often did not stay on the job long, quitting to find more interesting work. Ford solved this problem in 1914 by introducing a $5-per-day minimum wage—twice what factory workers were typically earning at the time—and cutting the working day to just eight hours. This announcement made Ford's workers the best paid in the country, and it caused such a sensation that police had to be called to control crowds who massed outside the Ford factory to apply for work. Ford also built hospitals and schools for his workers and even introduced a profit-sharing plan. Critics worried that Ford was focusing too much on workers' welfare rather than profits: the *Wall Street Journal* suggested that actions like his were an "economic crime."

Ford could defy the critics because his firm was spectacularly successful. His attention to workers actually increased his success, and not only in car production. He never forgot the tedium and hard labor of farmwork. Around the time he developed the Model T, he started tinkering with car parts to see if he could make what he called an "automobile plow," a combination of horse-drawn plow and automobile. The result of this effort was a tractor, the Fordson Model F, introduced in 1917, which did for agriculture what the Model T Ford did for driving. By using mass production methods, Ford cut the original price of the

The Model T Ford

"I will build a motor car for the great multitude, constructed of the best materials, by the best men to be hired, after the simplest designs that modern engineering can devise, so low in price that no man making a good salary will be unable to own one and enjoy with his family the blessing of hours of pleasure in God's great open spaces."

—Henry Ford

No one could accuse Henry Ford of breaking his promises. When the Model T Ford went on sale in 1908, it achieved all these aims. Originally priced at $850, the touring model was well beyond the reach of most potential customers. However, Ford's innovative assembly line and his highly motivated workforce soon slashed the cost of making a Model T, and Ford passed the savings on to his customers. By 1913, he had cut the price to $600; two years later it had dropped to $440, and by 1925 it was just $290, about a third of its original cost. At that price, even average families could buy a car—and with it they gained freedom. The Model T Ford even earned an affectionate nickname, the "Tin Lizzie," just like a member of the family.

Price of the Model T Ford (Touring Model): 1909 to 1927*

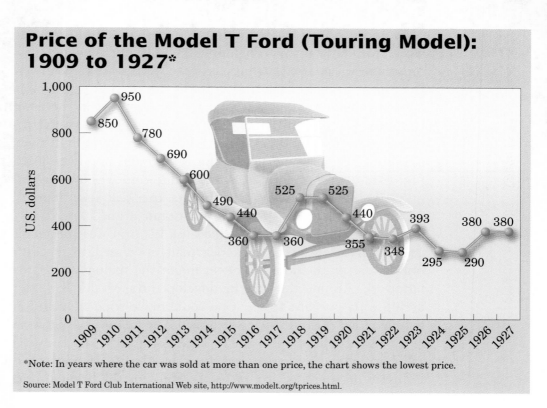

*Note: In years where the car was sold at more than one price, the chart shows the lowest price.

Source: Model T Ford Club International Web site, http://www.modelt.org/tprices.html.

Mass production brought other benefits. Since each worker was making only one piece of a car and all the pieces had to fit together, Ford had to use standardized components. Another of Ford's innovations had been to open dealerships where his cars could be sold; by 1912, he had 7,000 of these across the United States. Mass production made repair easier for dealers because they could use standardized spares. The process of putting a car together also had to be simple so cars could be assembled as quickly as possible. In 1912, it took the Ford workers about half a day (12 hours) to build each vehicle; two years later, they could make a car in 90 minutes. Customers also benefited because they could make basic repairs, such as changing tires, themselves.

The Model T's brilliance is difficult to appreciate in the 21st century. It is almost inconceivable today that a company could offer for sale a car that remained almost unchanged for 20 years and still sell 15 million. Between 1908 and 1927, more than half the cars sold in the United Sates were Model T Fords. When Henry Ford constructed his giant River Rouge plant in the 1920s, he did not have to borrow so much as a cent to do it: all the money he needed for this ambitious plan came from profits earned by the Model T.

Ford's idea of an affordable car was gradually copied by other manufacturers. The German carmaker Volkswagen even took its name from the idea; Volkswagen is German for "people's car." Basic, functional, and highly affordable, the car remains just as popular in the modern world as it was in Ford's own time a century ago.

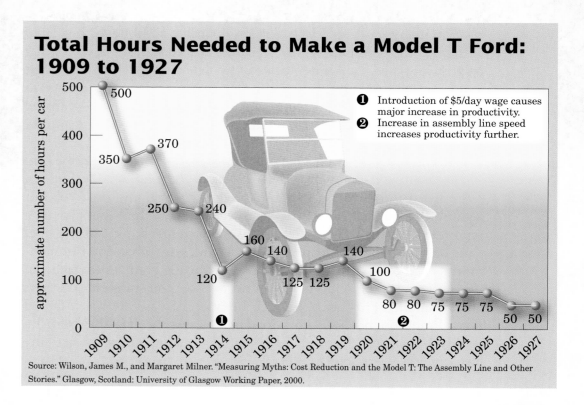

Total Hours Needed to Make a Model T Ford: 1909 to 1927

approximate number of hours per car

❶ Introduction of $5/day wage causes major increase in productivity.
❷ Increase in assembly line speed increases productivity further.

500, 350, 370, 250, 240, 120, 160, 140, 125, 125, 140, 100, 80, 80, 75, 75, 75, 50, 50

1909 1910 1911 1912 1913 1914 1915 1916 1917 1918 1919 1920 1921 1922 1923 1924 1925 1926 1927

Source: Wilson, James M., and Margaret Milner. "Measuring Myths: Cost Reduction and the Model T: The Assembly Line and Other Stories." Glasgow, Scotland: University of Glasgow Working Paper, 2000.

A Model T from 1916.

tractor from $750 to just $395. Eventually he sold 750,000 Fordson tractors and captured three-quarters of the U.S. market.

LIGHTNING ROD

With his company firmly established, Ford had time for other interests, including politics. He had strong opinions and never hesitated to express them. When World War I began in 1914, Ford, who was a pacifist, opposed the United States' involvement. In 1916 he rented a huge ocean liner, nicknamed the "Peace Ship," to sail to Europe in hopes of persuading the governments there to end the war—earning widespread ridicule. In 1918, he ran for the Senate in Michigan, but was narrowly defeated.

Later, Ford's views prompted the *Chicago Tribune* to describe him as an "anarchist enemy of the nation" and an "ignorant idealist." In 1919 his attorney persuaded him to bring a $1 million libel suit against the newspaper. Three years earlier, Ford had quipped to a reporter from the *Tribune*, "History is more or less bunk." In the libel case, defense attorneys leaped on the remark with what one commentator described as "sadistic pleasure." They made Ford appear to be an ignorant fool by proving he could not answer the simplest questions about American history. Although he won the case, he was awarded not $1 million but six cents in damages. As he left the trial, he resolved to open a Henry Ford Museum "to show just what actually happened in years gone by."

Although Ford's views were controversial, they were generally conservative. He set up a "Sociological Department" at his factory to discourage his well-paid workers from wasting their money on things he disliked. If workers expected to earn their $5 per day, they had to have their homes inspected and prove they were sober and clean-living. In other ways, Ford was extremely progressive. In the 1920s and 1930s, Ford's were among the only factories in the country to offer black workers equal opportunities.

FORD AND ANTI-SEMITISM

In 1918, Ford had used his wealth to buy a newspaper, the *Dearborn Independent*, so he could promote his political views more easily. This got him into trouble when he started attacking Jews, who he claimed had been responsible for financing World War I. In 1920, his newspaper columns were published as a book called *The International Jew: The World's Foremost Problem*, which offended many people. He issued a public apology in 1927.

Ford's apparent anti-Semitism led some to suggest links with Adolf Hitler, the brutal dictator who led Germany from 1934 to 1945. Ford's writings against the Jews were widely read in Germany in the 1920s and some historians have speculated that they may have aided Hitler's rise to power. It is known that Hitler greatly admired Ford and kept a pho-

Henry Ford's "Peace Ship" on its journey to Europe in 1916.

River Rouge

The enormous car plant Henry Ford created at River Rouge, Michigan, in the 1920s was a testament to his towering ambition. With 81,000 on the payroll and a total floor space of nearly 7 million square feet, the plant—not surprisingly—cost almost $270 million. This was a phenomenal sum early in the 20th century and prompted outrage among Ford's original shareholders: they feared such a massive investment would seriously dent the company's profits. Henry Ford had to buy out their shares personally so he could continue with his expensive plan.

Ford's idea was to build the world's largest car plant, so he could cut the price of the Model T still further. His philosophy of car manufacturing was "Reduce the price, extend the operations, and improve the article." However, he soon realized he would also need to lower the cost of raw materials. That prompted moves to make River Rouge entirely self-sufficient. Thus, the plant burned coal from Ford's own mines in Kentucky, and it used steel manufactured from his iron mines in Michigan, wood from his forests in Minnesota, rubber from the company's plantations in Brazil, and windows made in River Rouge's own glassworks. Although it was a logical idea for making cars less expensively, maintaining huge plants like River Rouge meant the company could not respond as quickly to consumer demand, a problem that hastened the Ford Motor Company's decline in the 1930s.

tograph of him on his wall. He even awarded Ford a medal for building a factory in Berlin to make trucks and tanks for the German military.

TYRANT

Although his cars and tractors were changing the face of America, Ford was nostalgic for an earlier and simpler time. He financed a radio show in which rural stories were read on air to his workers. In 1933, he expanded his museum into a historical theme park and school named Greenfield Village. Built in the style of a rural town, it incorporated reproductions of 100 buildings of major historical significance and has been a popular educational and tourist attraction ever since. Ford wanted to preserve the past, but not live in it. As he wrote in his autobiography, "What is past is useful only as it suggests ways and means for progress."

As he grew older, Ford became more set in his ways. Although his son Edsel became the nominal president of his company in 1919, Henry Ford still made all the key decisions, frequently overturning those made

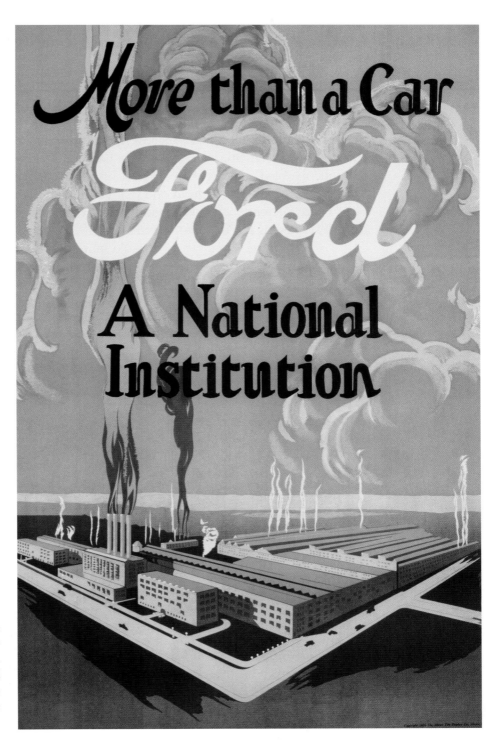

An advertisement for Ford Motor Company from 1923 emphasized the company's importance to the U.S. economy.

TIME LINE

1863	1869	1888	1896	1899
Henry Ford born in Greenfield, Michigan.	Ford begins an apprenticeship as a mechanic.	Ford works as an engineer in Detroit.	Ford builds his first automobile.	Ford forms the Detroit Automobile Company.

Henry Ford photographed in 1930.

by Edsel. Ford's extraordinary success made him reluctant to change, but the world was changing. Even when the General Motors Company began to challenge Ford's market share, Ford continued to make only one car, the Model T, never altering it from one year to the next. Ford's view was simple: "There is a tendency to keep monkeying with styles and to spoil a good thing by changing it." Toward the end of the 1920s, however, after Ford had sold more than 15 million Model Ts, he introduced a different car—named Model A, like the first vehicle he produced in 1903—to replace it. In the 1930s, he finally joined the rest of the car-making business and introduced a new model each year.

Not only car manufacturing had moved on—workers' expectations had also changed. By the 1930s, Ford was paying less than the industry average, and his workers were demanding to join unions. Ford resisted, turning his Sociological Department into a bullying police force called the Service Department. Run by a former boxer, it started cracking down on union organizers. In 1937, when the Service Department clashed with union officials at Ford's giant River Rouge plant, a small riot followed and 16 people were injured. After a major legal action and a strike at River Rouge, Ford was forced to choose between shutting down the factory and allowing in the unions. Grudgingly, he relented.

TIME LINE

1902	1908	1917	1945	1947
Ford starts the Ford Motor Company.	Ford launches the Model T.	Ford launches the Fordson Model F tractor.	Ford retires from Ford Motor Company.	Ford dies.

LATER YEARS

When Edsel Ford died of cancer in 1943, at only 49 years of age, the presidency of the Ford Motor Company reverted to his deeply saddened father. As the company's fortunes continued to decline, however, people questioned whether Henry Ford was the best person to run the company he had founded. The stubbornness and determination that had made him one of the greatest industrialists of all time now took his company to the brink of ruin. During the late 1930s, the company had been losing an estimated $1 million every day. World War II brought a respite. Although Ford again opposed the United States' involvement, he saw no contradiction in helping the government to manufacture thousands of warplanes at a huge new plant built specifically for the purpose.

When the war ended in 1945, Ford's reign ended, too. Control passed to his grandson, Henry Ford II; and Henry Ford, the founder, receded into the background. He died two years later, on April 7, 1947, in Dearborn, leaving a personal estate estimated at nearly $1 billion.

AFTER FORD

Only 28 when he took charge, Henry Ford II rapidly restored the Ford Motor Company's fortunes. He brought in new managers, modernized the business, and turned it into a global organization with bases in more than 30 countries. Although his grandfather had insisted on keeping control of the company in the family and private, Henry Ford II took Ford Motor public in 1956. This stock offering, which raised $650 million, was the largest stock offering in the United States to that time. Two years later, the company launched a radical new model, named the Edsel as a tribute to the late Edsel Ford. Despite being technically innovative and heavily advertised, the Edsel was a flop and the company was forced to scrap it within three years. This episode cost the company almost half the money it had made through the stock offering. Ford quickly recovered, however, remaining one of the world's largest industrial corporations into the 21st century.

In 2004, Ford Mustangs are assembled at a new, state-of-the-art plant in Flat Rock, Michigan, featuring 380 manufacturing robots.

When Henry Ford was born, most Americans lived rural lives, many on farms similar to the one where he had grown up. By the time he died, most lived in cities and many worked in factories. Ford was an engine of that change. He built highly mechanized factories where people could work, developed tractors that reduced the need for farm labor, created new jobs for workers in factories, and manufactured cars that almost every working family could afford, making Americans mobile in a way unimaginable to prior generations. It is often said that Ford's lasting achievement was in helping to invent a middle class of people with both money and the time to enjoy it.

—Chris Woodford

Further Reading

Books

Ford, Henry. *My Life and Work*. Whitefish, MT: Kessinger, 2004.

Sutton, Richard. *Eyewitness: Car*. New York: Dorling Kindersley, 2005.

Watts, Steven. *The People's Tycoon: Henry Ford and the American Century*. New York: Knopf, 2005.

Web sites

Henry Ford Museum
    Henry Ford's museum of American life.
    http://www.thehenryford.org
Museum of Automobile History
    The world's largest collection of car history exhibits.
    http://www.themuseumofautomobilehistory.com/

See also: Benz, Karl; Deere, John; Edison, Thomas; Transportation; Whitney, Eli.

BENJAMIN FRANKLIN

Prolific inventor

1706–1790

Benjamin Franklin filled many positions during his life: a founding father of the United States, a diplomat and statesman, a printer and journalist, and a public servant. He was also a scientist with a curious and creative mind who developed a variety of inventions. Like his politics and his public service, his life as an inventor was focused on improving people's lives.

EARLY YEARS

Benjamin Franklin, one of the most famous Americans of the 18th century, was born on January 17, 1706, in Boston, Massachusetts. Josiah
Franklin, his father, was a soap- and candle-maker; what little money
Josiah earned was used to support his 17 children, of whom Benjamin was
the 15th. Benjamin attended grammar school between the ages of 8 and
10, then left to work in his father's shop, where he helped by cutting can-

*Benjamin
Franklin; oil
painting by David
Martin, 1767.*

dlewicks, molding wax, and running errands. Around this time, he also developed a passion for swimming and created his first invention: a set of swim fins that could propel him quickly through the water.

At age 12, he became an apprentice to his brother James, a printer in Boston. From James, Franklin learned the printing trade that would soon make him famous. During the day he helped James print his newspaper, the *New England Courant*; in the evening he studied English. He would read essays from great books and then try to rewrite them in his own words; he gradually mastered an impressive style. One day, for a joke, Franklin started writing funny letters to his brother's paper and submitting them under the pseudonym Mrs. Silence Dogood. Although the letters were popular with readers, James was furious when he found out who had written them. The brothers had a falling out and Benjamin decided to leave Boston.

> If you would not be forgotten
> As soon as you are dead and rotten,
> Either write things worth the reading,
> Or do things worth the writing.
>
> —Benjamin Franklin

TRAVELER'S TALES

After a brief period in New York City, the 17-year-old Benjamin Franklin traveled to Philadelphia. He found work as a printer and made friends with William Keith, the governor of Pennsylvania. Keith advised Franklin to go to London to finish training as a printer, even offering to pay his expenses. Franklin accepted the offer, but Keith's money was never forthcoming. When he arrived in London, he was stranded and had to use his printing skills to find work to support himself. After spending about eighteen months in London, where he met many famous writers and publishers, he had saved enough money to return to Philadelphia. There, in 1728, he started his own print shop—he was only 22.

Franklin worked hard at his printing business, becoming wealthy and successful. He was the official printer for Philadelphia, New Jersey, Delaware, and Maryland, but he became better known for an amusing book, *Poor Richard's Almanac*, printed annually from 1732 to 1757. This almanac combined a calendar and weather forecasts with amusing stories and sayings. Many of the proverbs of this book are still popular, including "Haste makes waste" and "Early to bed, early to rise, makes a man healthy, wealthy, and wise."

EARLY INVENTIONS

Some of Franklin's best-known inventions date from this time. In 1740, he began to think about the way people heated their homes by burning fuel in large open fireplaces. After some experimenting, he invented what became known as the Franklin stove: a sturdy cast-iron basket with a large, open front that drew in more air, burned less fuel more efficiently, and gave off less smoke and more heat than an open-hearth fireplace. Often made in a freestanding form, it could be placed in the middle of a room to warm the area more evenly. Franklin's stove was also safer because the metal basket held the burning fuel more securely.

Most of Franklin's inventions were designed to overcome everyday nuisances. Some, such as his invention of bifocal eyeglasses, were inspired by his own experiences. When people pass the age of 40, their eyesight changes in a way that makes seeing close objects difficult. As Franklin already wore eyeglasses, he was confronted with the inconvenience of needing two different pairs, one for seeing close-up and another for look-

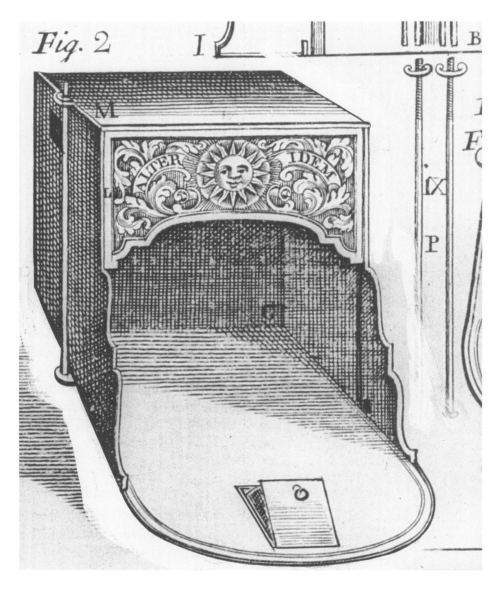

Undated illustration of an original Franklin stove.

ing into the distance. This inspired Franklin to invent bifocals: he took his two pairs of eyeglasses, cut the lenses in half, and then joined the opposite halves together to create a single pair for seeing close-up when he looked down, or into the distance when he looked up.

PUBLIC SERVICE

Franklin had many good ideas, but not all of them were for inventions—some were ideas for improving society. From the 1730s on, he devoted much of his time to the public services of Philadelphia. He had the idea to pay for a "city watch," as it was then known, which became the city's first police force. He campaigned successfully for improvements to street cleaning, paving, and lighting. He helped found a hospital. He also organized the first volunteer fire company in America. His love of knowledge and books prompted him to establish what is widely believed to have been the country's first public library; it housed books, a museum of scientific instruments, and a collection of mounted and stuffed wildlife specimens, including a pelican. Later, Franklin founded an academy, opened in 1751, that eventually became the University of Pennsylvania.

Many of his civic improvements were concerned with the post office. Between 1737 and 1753, Franklin served as deputy postmaster to Philadelphia. He quickly identified many problems with the postal service and put improvements in place. Postal fees were charged according to the weight of each item and how far it had to be carried from the sender to the receiver. However, no one knew the distances between cities at that time, so disputes often arose about how much to charge. Franklin solved this problem with another of his inventions, the odometer. This was a simple measuring device attached to the wheels of a mail carriage. As the carriage rolled along, the odometer's dials spun, recording the exact distance traveled. With that information, a precise mailing fee could be calculated.

SCIENTIFIC DISCOVERIES

By the end of the 1740s, Franklin had earned enough from his printing business to retire. He bought a large 300-acre farm near Burlington, New Jersey, and began to devote himself to the study of electricity. Almost nothing was then known about the subject: what electricity was, how it flowed, and how it could be stored were all questions waiting to be answered. Franklin played a major part in unraveling the mystery. Electrical research earned him world fame in the 1750s, when he published his results in a series of pamphlets (later collected into a book, *Experiments and Observations on Electricity*). Soon translated into French, Italian, and German, the pamphlets helped European scientists begin to regard electricity as a usable form of energy rather than a curiosity of the natural

Inventing for the Common Good

Lightning rods, busybodies, stoves, eyeglasses, and swim fins—
these are just some of the devices Benjamin Franklin brought into
being during his long and highly productive career as an inventor.
Unlike many inventors before and since, Franklin was not motivated by
money. His printing business had made him wealthy, so he did not need
his inventions to support him or add to his wealth. Some of his inventions would have earned him a fortune, yet he steadfastly refused to
patent any of them.

When he developed the Franklin stove, the governor of Pennsylvania
offered him a patent on the invention so he could make money from it,
but he refused. He wanted the stove to be manufactured inexpensively to
make it affordable to as many people as possible. Commenting on this in
his autobiography, he said, "As we enjoy great advantages from the inventions of others, we
should be glad of an
opportunity to serve
others by any inven-
tion of ours; and this
we should do freely and
generously."

*Undated illustration of
Franklin and one of his
inventions, bifocal lenses.*

An illustration by Currier and Ives depicts Franklin and his son William flying a kite during a thunderstorm.

world. Franklin's work was an immense breakthrough (although its fundamental importance was not immediately realized) and did have an immediate practical benefit— it led to one of his most famous inventions: the lightning rod.

In 1752, in one of the most famous scientific experiments in history, Franklin went out in a thunderstorm to fly his kite. After several years studying electricity, he had arrived at the theory that lightning and electricity were somehow connected; all he had to do was prove it. With his kite dipping and rattling high in the air, he tied a metal key to the end of the kite's long, wet string. When he moved one of his knuckles toward the key, he felt a small tingle: electricity was traveling down the string from the storm, through his finger, and running through his body to ground. Franklin was lucky not to have been killed.

The experiment demonstrated that a lightning bolt is a massive discharge of electricity from the sky to the ground. Such an immense amount of electrical energy can kill people instantly, set buildings on fire, and make trees explode by boiling the liquids inside them dry. However, Franklin saw that lightning could be tamed by creating a more direct path for the electricity to follow. He invented the lightning rod: a strip of metal that runs down the side of a high building, carrying the electricity in lightning safely to ground.

The kite experiment confirmed Franklin's theory that electricity was a kind of fluid that could flow from place to place. A French scientist,

Charles Du Fay (1698–1739), had explored this idea 20 years earlier and was convinced that electricity was of two kinds. However, through his experiments, Franklin came to a very different conclusion: electricity was of a single kind but could flow in different directions. If it flowed into an object, that object would effectively gain electricity: Franklin described this as a negative charge. If it flowed out of an object, that object would lose electricity, gaining a positive charge. The movement of positive and negative charges around a circuit eventually came to be known as electric current.

Increasing the understanding of electricity was one of Franklin's most famous contributions to science. Later, he carried out important experiments on the way heat flows. He was one of the first to understand how evaporation can cool liquids and solids. While wearing a wet shirt on a hot day, he noticed that he stayed cool because water evaporating from the shirt removed heat from his body. Thus, he concluded: "One may see the possibility of freezing a man to death even on a warm summer's day."

A cartoon by Franklin from around 1770 encouraged American colonists to band together against British rule.

In the 1740s, Franklin observed that stormy weather does not always travel in the same direction as the wind. This observation helped to improve the accuracy of weather forecasts. Around thirty years later, Franklin investigated another important influence on the weather: the Gulf Stream, a warm current of water that flows north and west through the Atlantic Ocean off the East Coast of the United States. Although Franklin did not discover the Gulf Stream (sailors had known of it since

Understanding Electricity

Franklin's experiments in electricity were ground-breaking because they demonstrated that electricity could be made to move from one place to another in a controlled way. This realization marked the birth of electric current. Unlike previous theories of electricity, which were only partly correct, the theory of positive and negative charges was a more complete explanation of what electricity was actually doing. Thus, Franklin's biggest contribution was to put the science of electricity on the correct conceptual path, which helped humans swiftly move forward in their understanding and use of this elemental force.

Two Italian scientists, Luigi Galvani (1737–1798) and Alessandro Volta (1745–1827), read Franklin's writings on electricity. In 1800, Volta ushered in the modern age of electricity when he became the first person to make a practical electric battery. Other pioneers, including the English scientist Michael Faraday (1791–1867) and the prolific U.S. inventor Thomas Edison (1847–1931), then turned electricity into a useful everyday source of power. Franklin's original work was crucial to these advances.

In 1898, just over a century after Franklin's death, his research helped scientists discover the electron, the tiny particle inside atoms that carries electricity around circuits. In fact, the concept of positive and negative charge, which Franklin delineated, is central to the modern scientific understanding of atoms. Robert Millikan (1868–1953), an American physicist who won a Nobel Prize for his pioneering work with electrons, described Franklin's kite experiment as "probably the most fundamental thing ever done in the field of electricity."

Nobel-winning scientist Robert Millikan in his laboratory at the California Institute of Technology (undated photograph).

the 1500s), he measured its temperature, speed, and depth and also drew the earliest maps of it.

STATESMAN AND FOUNDING FATHER

Franklin had already achieved much as a printer, public servant, scientist, and inventor. In the 1750s, a new phase of his life began that would make him a world-famous statesman. In 1754, conflict between France and Britain spread to their colonies in North America, in what is referred to as the French and Indian War. French soldiers in Canada joined forces with Native Americans and began raiding Pennsylvania, then still a British colony. In 1767, Franklin was dispatched to London to raise support for the people of Pennsylvania from the king of England. He remained there for eight years, during which time he met and shared his ideas with some of the most famous scientists in Europe.

When he returned to Philadelphia in 1775, the opening battles of the War of Independence had already been fought. Franklin was now 69 and was widely respected for his wisdom. He was one of the five men who drafted and signed the Declaration of Independence in 1776. As the war progressed, Franklin spent time in France, seeking backing for America in the fight against Britain.

FINAL YEARS

When Benjamin Franklin returned again to Philadelphia, on September 14, 1785, practically the entire city turned out to welcome him home. Cannons were fired, bells were rung, and the celebration lasted a week. Although he was nearly 80 and somewhat frail, he remained an active statesman. For the next three years, he served as president of the Pennsylvania Executive Council. He also helped to draw up and sign the Constitution of the United States in 1787. One of his last acts was to sign a petition to the U.S. Congress on February 12, 1790, recommending the abolition of slavery.

TIME LINE

1706	1718	1728	1732	1737–1753	1740
Benjamin Franklin born in Boston, Massachusetts.	Franklin becomes a printer's apprentice in Boston.	Franklin starts his own printing business in Philadelphia.	Franklin begins publishing *Poor Richard's Almanac.*	Franklin serves as the deputy postmaster to Philadelphia.	Franklin invents the Franklin stove.

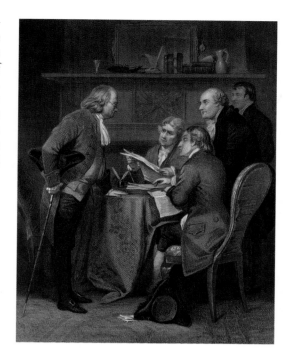

An undated illustration of the drafting of the Declaration of Independence. From left: Franklin, Thomas Jefferson, John Adams, Robert Livingston, and Roger Sherman.

In poor health, Franklin spent his last years at home, where he devised ingenious inventions to make life easier. He constructed a rocking chair with curved rollers attached to the legs so he could move back and forth. His writing chair had a large piece of wood attached to one arm so he did not have to sit at a desk to work. Unable to reach books on high shelves, he invented a mechanical arm to help him lift them down. For the last year of his life, Franklin was bedridden and in constant pain. He devised a pulley system to lock and unlock his door while he was lying in his bed. Reputedly, he also invented the famous "busybody," three carefully positioned mirrors that allowed him to see visitors at his front door without getting up. Franklin died on April 17, 1790, at age 84.

Franklin's highly practical inventions made daily life easier for many people: his Franklin stove was commonplace in American homes for more than one hundred years, and bifocal eyeglasses remain popular. His investigations into electricity helped to turn it from a mystery into a science that others could explore, laying the groundwork for later inventors such as Thomas Edison to turn electricity into a useful technology that would transform the world.

—Chris Woodford

TIME LINE

Late 1740s	1752	1776	1787	1790
Franklin retires from his printing business and studies electricity.	Franklin conducts his kite experiment.	Franklin helps write the Declaration of Independence.	Franklin helps write the U.S. Constitution.	Franklin dies.

Further Reading

Books

Fleming, Candace. *Ben Franklin's Almanac: Being a True Account of the Good Gentleman's Life.* New York: Atheneum/Anne Schwarz, 2003.

Franklin Institute. *The Ben Franklin Book of Easy and Incredible Experiments.* San Francisco: Jossey-Bass, 1995.

The Old Farmer's Almanac editors. *Ben Franklin's Almanac of Wit, Wisdom, and Practical Advice: Useful Tips and Fascinating Facts for Every Day of the Year.* Dublin, NH: Yankee, 2003.

TIME editors. TIME *for Kids: Benjamin Franklin: A Man of Many Talents.* New York: HarperTrophy, 2005.

Web sites

Benjamin Franklin: An Extraordinary Life. An Electric Mind.
An interactive PBS Web site dedicated to Franklin's life.
http://www.pbs.org/benfranklin/

Ben's Guide to U.S. Government for Kids
An introduction to federal government, guided by Benjamin Franklin.
http://bensguide.gpo.gov/

Franklin Institute
Information, games, and activities from Philadelphia's science and technology museum.
http://wwwz.fi.edu/

See also: Edison, Thomas; Energy and Power; Faraday, Michael; Science, Technology, and Mathematics; Volta, Alessandro.

GEORGE FULLER

Inventor of the skyscraper
1851–1900

George Fuller is credited with inventing the modern skyscraper: in particular, with developing the steel cage at the heart of these tall buildings. His construction firm, the George A. Fuller Company, was responsible for some of the most famous skyscrapers and national monuments in the United States. Fuller, however, did not live long enough to see the advent of the modern skyline—what some call the "vertical metropolis"—that he helped create.

EARLY YEARS

George Fuller was born in Templeton, Massachusetts, in 1851. After graduating from Andover College, he attended the Boston School of Technology. He later began work at a famous Boston architectural firm, Peabody & Stearns, quickly advancing from drawing building plans to achieving partnership. At the age of 25, he took charge of the company's New York office.

Fuller took a particular interest in the load-bearing capacity of buildings—how much weight various parts of a building could support. He moved to Chicago and founded a construction company in 1882.

Constructed at the close of the 19th century, the Monadnock Building in Chicago is the world's tallest building to have load-bearing walls.

How to Build a Skyscraper

Each skyscraper is unique. Every design is based on a number of distinct factors, including the building's location and the surrounding environment. Even the weather plays a role, since, for tall buildings, the effect of wind on the structure is a vital consideration. Most skyscrapers, however, are constructed along similar lines.

First, the foundation is built. Builders dig into the ground, sometimes for hundreds of feet, until they reach bedrock, the layer of solid rock that lies beneath the soil. Footings, or large holes, are drilled into the bedrock. Columns made of steel or reinforced concrete are inserted into the footings, forming the base of the building. Then concrete is poured to create the bottom floor. This is sometimes called the substructure.

Construction continues up from the support beams, creating the vertical support for the building. Horizontal beams connect to these vertical supports and hold the building together. In many designs, the horizontal beams form the basis for the floors of the building. These two steps often happen simultaneously—with the horizontal beams of the lower floors being put in place at the same time the vertical beams are being added to the top. Sometimes reinforced concrete is used to bolster the strength of the building. Often this latticework of beams is called the superstructure.

Once the superstructure is complete, the exterior is built. The substructure and superstructure carry the weight of the building, so the exterior is primarily ornamental and can therefore be constructed from a variety of materials, including glass, metal, and stone. When the roof is put in place and the interior elements, such as plumbing, lighting, and electrical features, have been added, the skyscraper is complete.

A skyscraper under construction in Ukraine in 2006.

The George A. Fuller Company, often called simply the Fuller Company, would become one of the most successful construction companies in the United States.

THE STEEL CAGE

Several inventions paved the way for Fuller's skyscraper—most notably, steel. An Englishman, Henry Bessemer (1813–1898), discovered how to mass-produce steel efficiently and inexpensively in the 1850s. By the 1860s, the Bessemer process was beginning to revolutionize American manufacturing and construction. Elevators, invented in 1852 by Elisha Otis (1811–1861), were another vital innovation, as were central heating and electric power.

At the end of the 19th century, buildings were rarely more than four or five stories. These buildings were constructed with load-bearing walls, meaning that the walls themselves had to be thick enough to support the weight of the building, which imposed certain limits on height. For instance, the Monadnock Building in Chicago, one of the tallest buildings constructed with load-bearing walls, reached an impressive 16 stories, but its walls were six feet thick at the base.

In 1889 Fuller built the first skyscraper without load-bearing walls, the Tacoma Building in Chicago. He used, instead, a steel cage that could support the entire weight of the building. This was, by far, the most significant engineering development in the construction of tall buildings at that time. The outer walls were no longer part of the central strength of the building; they simply kept out the wind and rain and provided a shell or ornamental facade. With steel-cage construction, walls could be much thinner, allowing for more room inside the buildings. No longer hampered by thick masonry walls, and bolstered by the strength of steel, buildings could now reach ever higher toward the sky (see box, How to Build a Skyscraper).

> The unique triangular shape of the Flatiron Building created strong wind currents that were known to blow women's skirts in the air. Police would have to chase gawkers from the 23rd Street area, leading to the saying "23 skidoo."

TIME LINE

1851	1876	1882	1889	1893	1900
George Fuller born in Templeton, Massachusetts.	Fuller takes charge of Peabody & Stearns's New York office.	Fuller moves to Chicago and founds the George A. Fuller Company.	Fuller builds the Tacoma Building in Chicago.	Fuller Company builds many structures for the Chicago world's fair.	Fuller dies.

The triangular Flatiron Building, pictured center, in downtown Manhattan.

World's Ten Tallest Buildings, 2006

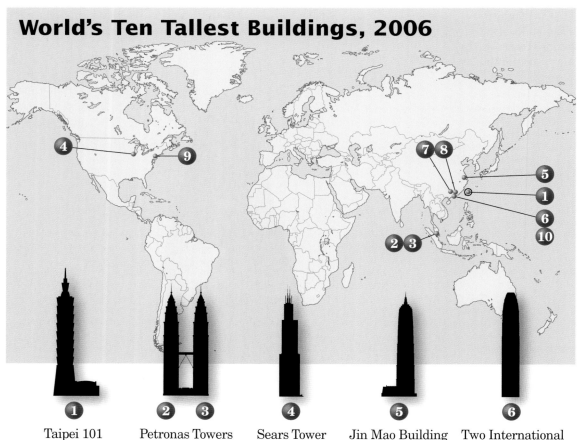

1

Taipei 101
Taipei, Taiwan
Year: 2004
Stories: 101
Height (m): 509
Height (ft.): 1,670

2　**3**

Petronas Towers
1 and 2
Kuala Lumpur,
Malaysia
Year: 1998
Stories: 88
Height (m): 452
Height (ft.): 1,483

4

Sears Tower
Chicago,
United States
Year: 1974
Stories: 110
Height (m): 442
Height (ft.): 1,450

5

Jin Mao Building
Shanghai, China
Year: 1999
Stories: 88
Height (m): 421
Height (ft.): 1,380

6

Two International
Finance Center
Hong Kong
Year: 2003
Stories: 88
Height (m): 415
Height (ft.): 1,362

7

CITIC Plaza
Guangzhou, China
Year: 1996
Stories: 80
Height (m): 391
Height (ft.): 1,283

8

Shun Hing
Square
Shenzhen, China
Year: 1996
Stories: 69
Height (m): 394
Height (ft.): 1,260

9

Empire State
Building
New York,
United States
Year:1931
Stories:102
Height (m): 381
Height (ft.): 1,250

10

Central Plaza
Hong Kong
Year: 1992
Stories: 78
Height (m): 374
Height (ft.): 1,227

George Fuller could not have constructed the Tacoma Building without the help of architects William Holabird (1854–1923) and Martin Roche (1853–1927), who often share the credit with Fuller for inventing the first skyscraper. Holabird and Roche were two of a number of architects who were considered practitioners of the Chicago school of architecture (sometimes called the First Chicago school). The Chicago school arose after the Great Fire of 1871; the steel cage with masonry facade became one of the significant features of the style.

NEW YORK'S FINEST

The George A. Fuller Company enjoyed much success after the Tacoma Building. Fuller himself oversaw the construction of more than two dozen of Chicago's tallest buildings, and the Fuller Company helped build many of the structures for the 1893 world's fair in Chicago.

In the late 1890s, Fuller and Chicago architect Daniel Burnham traveled to New York to build a new headquarters for the Fuller Company—New York's first skyscraper, in the very heart of Manhattan. The building was to be 20 stories high, built on a narrow triangular wedge of land flanked by Broadway, Fifth Avenue, and 22nd and 23rd

The Lincoln Memorial in Washington, D.C., built by the George Fuller Construction Company.

Streets. Originally called the Fuller Building, it is now known as the Flatiron Building. It remains New York's oldest surviving skyscraper. Unfortunately, Fuller did not live to see its completion; he died on December 14, 1900.

The George A. Fuller Company continued to grow into one of the country's leading contractors, responsible for the construction of the Lincoln Memorial, the National Archives, and countless commercial buildings in the United States, none of which would have been possible without Fuller. Fuller was among the first 19th-century engineers to transform the way that both architects and society itself viewed buildings. His contribution can still be seen in the ongoing race toward the sky, with buildings such as Chicago's Sears Tower, New York's Empire State Building, and Taiwan's Taipei 101, which, as the name suggests, is 101 stories.

—Laura Lambert

Further Reading

Book

Daly, Raymond C. *75 Years of Construction Pioneering; George A. Fuller Company, 1882–1957*. Erie, PA: Newcomen Society in North America, 1957.

Web sites

PBS.org: Building Big: All about Skyscrapers
    Educational Web site about skyscrapers and their history.
    http://www.pbs.org/wgbh/buildingbig/skyscraper/
Skyscraper Museum
    Online home of New York City's Skyscraper Museum.
    http://www.skyscraper.org/home_flash.htm

See also: Bessemer, Henry; Buildings and Materials; Otis, Elisha; Reno, Jesse.

R. BUCKMINSTER FULLER

Inventor of the geodesic dome

1895–1983

R. Buckminster Fuller was more than an inventor; to the many followers he attracted during his long life of lecturing, he was a true visionary. Fuller, who wrote 28 books on a wide range of subjects, made important contributions in the fields of engineering, architecture, and design. Fuller's best-known contribution to humanity may be his coining of the phrase "spaceship earth," to express what he saw as our common need to use technology wisely and avoid wasting our resources. Above all else, Fuller is known as the inventor of the geodesic dome, one of the great breakthroughs in design technology.

EARLY YEARS

Richard Buckminster Fuller was born July 12, 1895, in Milton, Massachusetts. The young Fuller learned about seamanship and became interested in the ways people solve the various practical problems their surroundings present. One early solution to a personal obstacle came when Fuller received eyeglasses at the age of four. After seeing the world as a series of blurs up to that point, he was suddenly, as he recounted years later, "filled with wonder at the beauty of the world and . . . never lost delight in it." He also began to demonstrate his design talent, making his first design—an octet truss (an interlocked series of triangles) from toothpicks and dried peas while in kindergarten. This shape would later become integral to his invention of the geodesic dome.

Fuller was born into a distinguished, well-to-do New England family. His father, also Richard Buckminster Fuller, was a successful Boston merchant whose aunt was the well-known writer and social reformer, Margaret Fuller (1810–1850), and whose great-uncle was Melville Fuller (1833–1910), who was appointed chief justice of the United States in 1888. Earlier generations of Fuller men had attended Harvard, and the young Richard Fuller, or "Bucky" as he was known, had no difficulty following in their footsteps. He entered Harvard in 1913 but was expelled twice in two years for "irresponsible behavior."

R. Buckminster Fuller photographed with some of his tetrahedrons in 1964.

After the second expulsion, Fuller never again pursued formal education.

Fuller served briefly as a radio operator in the U.S. Navy during World War I. He married Anne Hewlett in 1917 and worked at a variety of jobs. In the early 1920s he joined with his father-in-law, the architect James Monroe Hewlett, to form the Stockade Building System, which specialized in producing a type of reinforced, fibrous building block. That company failed in 1927.

That same year, Fuller came to the brink of suicide. His daughter Alexandra had died of pneumonia, his business had failed, and he felt that his life was in ruins. However, as he stood on the shores of an icy Lake Michigan in Chicago, Fuller recounts that he said to himself, "You do not have the right to eliminate yourself; you do not belong to you. You belong to the universe." From that moment Fuller committed his life to improving the living conditions of humanity.

THE DYMAXION CONCEPT

Feeling that a poor living environment had contributed to his daughter's death, Fuller immersed himself in designing improved housing for the masses. He founded the 4D Company that year and developed the Dymaxion House—the name was derived from dynamic-maximum (efficiency) plus ion, or power. Circular in design (to prevent heat loss), the domicile was sometimes called a "house on a pole" because it consisted of a lightweight material hung over a central pole. The Dymaxion House was to be heated by solar energy. Fifty feet in diameter (15 m) and weighing approximately six thousand pounds, the house was designed to rotate

An interior view of Fuller's Dymaxion House, photographed in 1941.

Fuller explains his Dymaxion AirOcean World Map, which he argued was a better representation of the world than a traditional map.

to follow (or, if desired, avoid) the sun. Fuller sought in every detail to maximize efficiency, as he did in all his designs. The Dymaxion House operated according to strict recycling principles: it contained a water filtering and reuse system, a solid-waste composting system, a low-flow showerhead called a "fogger," and mattresses filled with air.

Fuller's plan was to assemble thousands of these houses in a factory, to be delivered complete to people all over the world. The Dymaxion House drew plenty of attention but never gained acceptance as a practical alternative to existing home designs, despite Fuller's insistence that the world would be made better by adopting his designs. Several years after developing the initial concept, Fuller founded the Dymaxion

Dwelling Machines Company in 1944, in Wichita, Kansas. Hoping to mass-produce the 1,000-square-foot Dymaxion House (now called the Wichita House), Fuller considered his design an ideal answer to the post-war housing shortage he saw coming. Financial backers were not convinced, however, and the company soon failed, despite having received several orders. Many observers today feel that the housing crisis of the 1970s, along with the problems this crisis brought to American cities, might have been less acute had America in 1945 embarked on a course in keeping with Fuller's vision of suitable, inexpensive, widely available dwellings.

If his housing design raised eyebrows, Fuller's Dymaxion Car was even more controversial. Developed in 1933, this car featured a three-wheel, front-wheel-drive design that resembled an airplane body. With a Ford 90-horsepower V-8 engine, the Dymaxion Car was designed to hold 12 passengers and travel 120 miles per hour (193 km/h) while getting 40 miles per gallon (17 km/l). Only three of these cars were produced between 1933 and 1936. Research on the project ended abruptly after one of the cars crashed, leaving the driver dead. It later emerged that the crash was the fault neither of the Dymaxion Car nor of the driver, but rather of the driver of another car. Yet the bad publicity that resulted from the crash, in combination with the car's unconventional appearance, doomed the Dymaxion Car's future.

Fuller continued developing the Dymaxion concept, however, creating in 1936 the Dymaxion Bathroom, which could provide a 10-minute bath using only a quart of water, and could give a massage. In 1946 he created a Dymaxion AirOcean World Map, a map in a complex cube-like shape that could be unfolded in various ways to show the relations among the continents and oceans of the world or flattened into a two-dimensional map. Fuller insisted that there was no one right way to fold the map and claimed it offered a truer representation of the planet while avoiding the cultural bias of north being "up" (and by implication superior) and south "down" (and thus inferior).

> I set about fifty-five years ago to see what a penniless, unknown human individual with a dependent wife and newborn child might be able to do effectively on behalf of all humanity.
>
> —R. Buckminster Fuller

THE GEODESIC DOME

Unsure what to do after the successive failures of his Dymaxion concepts, in 1948 Fuller took a teaching job at Black Mountain College, the experimental educational institution in North Carolina. At Black Mountain College, all of Fuller's work in maximizing efficiency in various objects and materials would culminate in his greatest invention, the geodesic dome. Since Fuller's first unveiling of the geodesic dome, thousands of these innovative structures have been built around the world.

How the Geodesic Dome Works

When Buckminster Fuller was in kindergarten, his teacher was astonished to see, among the simple boxes that students were constructing out of the toothpicks and dried peas she had given them as a project, something totally different: an octet truss. This interlocked triangle-shaped structure resembling a honeycomb would years later change the world. It would be the basis of Fuller's geodesic dome.

Buckminster Fuller would later write that geodesic meant "the most economical momentary relationship among a plurality of points and events." In fact, Fuller maintained that his idea for the geodesic structure of interlinked tetrahedrons was present everywhere in nature, such as in the eye's cornea. Believing that nature operates in the most efficient ways, he tried to copy the structures he saw around him, using mathematical principles to translate nature's forms for use by humans. The truth of his belief was quickly established: Fuller's domes proved capable of supporting themselves to a greater extent than any other structure in the history of design.

The geodesic dome becomes stronger, lighter, and cheaper per unit of volume as the size increases. For hundreds of years, humans had constructed domes based on an understanding that a sphere is an inherently efficient shape: it can enclose the greatest volume using the least surface, thereby making it capable of conserving heat and withstanding winds better than nonspherical structures. The geodesic dome, however, differs from the traditional dome in its interlocking, self-bracing pattern of triangles. This pattern makes possible the construction of a sphere of enormous (theoretically unlimited) size without losing any stability. As the size of a geodesic dome increases, the weight load becomes distributed throughout the entire structure.

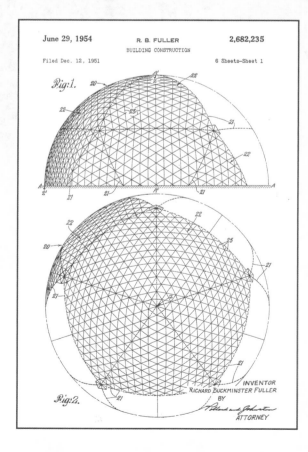

Illustrations from Fuller's 1954 patent.

An exterior photograph of the U.S. Pavilion, from the 1967 Montreal world's fair.

Fuller's dome is constructed of a series of connected triangles called tetrahedrons. His aim was to make a spherical enclosure that would encompass maximum space and achieve maximum strength using the minimum of materials. Having conceived the idea in 1947, he built his first sizable dome with his students at Black Mountain College the following year. Lightweight, resistant to wind, and quickly assembled, the geodesic dome could cover several acres of area with no internal supports.

In 1953 the Ford Motor Company commissioned Fuller to design a dome constructed of lightweight plastic and aluminum for the rotunda of its plant in Dearborn, Michigan. After the successful construction of the dome, which measured 93 feet (28 m) across, Fuller's reputation spread worldwide. More contracts followed, including a dome aircraft hangar for the U.S. Marine Corps in 1954 and "radomes" (enclosures for radar installations) for the Air Force DEW (Distant Early Warning) Line system across parts of Alaska and Canada in 1955. Fuller received a patent for the geodesic dome in 1954 (see box, How the Geodesic Dome Works).

FULLER'S LEGACY

Fuller left Black Mountain College in 1949 and formed Geodesics, Inc., in Forest Hills, New York. His work from that period was largely devot-

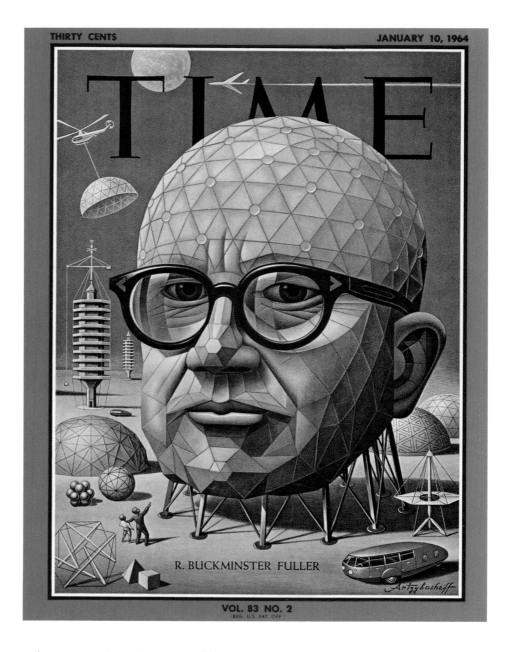

THIRTY CENTS

JANUARY 10, 1964

TIME

R. BUCKMINSTER FULLER

Artzybasheff

VOL. 83 NO. 2
(REG. U.S. PAT. OFF.)

ed to expanding the uses of his great invention. In 1959 he accepted a position as a research professor at Southern Illinois University in Carbondale, where he remained until 1970. There, with the resources of an entire building and a staff devoted to his cause, Fuller created ever-larger domes. One of Fuller's largest domes, the Houston Astrodome, was built in 1965. The U.S. Pavilion at the 1967 world's fair in Montreal featured a dome 200 feet (61 m) high and 250 feet (76 m) in diameter. It captured the attention of all who attended and became one of the symbols of the fair.

TIME LINE

1895	1927	1933	1944
Richard Buckminster Fuller born in Milton, Massachusetts.	Fuller founds the 4D Company.	Fuller develops the Dymaxion Car.	Fuller founds the Dymaxion Dwelling Machines Company.

Although Fuller's plan to construct a dome two miles (3 km) in diameter to cover a large portion of Manhattan Island never gained support (no one at the time shared Fuller's enthusiasm for such a large-scale "controlled environment" in the center of New York City), the Astrodome and the U.S. Pavilion are outstanding examples of the possibilities of these huge structures. More than two hundred thousand geodesic domes have been built as homes and shelters around the world.

R. Buckminster Fuller remains a unique figure in American culture and an inspiration to untold numbers of architects, designers, and environmentalists. Until his death in 1983, Fuller traveled and lectured widely, often holding audiences spellbound for four hours or more with extemporaneous insights. From the late 1960s through much of the 1970s, he was a hero among the student counterculture, who saw in him a visionary intent on improving the conditions of life for human beings. The Buckminster Fuller Foundation continues to promote Fuller's ideas

The Astrodome in Houston, Texas, designed by Fuller.

TIME LINE (continued)

1948	1954	1965	1983
Fuller builds the first geodesic dome.	Fuller receives a patent for the geodesic dome.	Fuller designs the Houston Astrodome.	Fuller dies.

and legacy, and to feature work by architects and designers who are following his lead. Fuller's *Operating Manual for Spaceship Earth* (1969) lays out his aims to use technology to achieve a more efficient, self-sustaining world.

—Paul Schellinger

Further Reading

Books

Baldwin, James T. *Buckyworks: Buckminster Fuller's Ideas Today.* New York: Wiley, 1996.

Pawley, Martin. *Buckminster Fuller.* New York: Taplinger, 1990.

Web site

Buckminster Fuller Institute
> A biography, design overview, and assessment of Fuller's legacy.
> http://www.bfi.org/

See also: Buildings and Materials.

INDEX

Page numbers in **boldface** type refer to article titles.
Page numbers in *italic* type refer to illustrations or other graphics.

PHOTOGRAPHIC CREDITS

Cover photos (clockwise from left): Miguel Angelo Silva/iStockphoto; Bill Pugliano/Getty Images; NASA; Science Museum. Apple, Inc: 346; **Associated Press:** 339; **Corbis:** Pedro Armestre/epa 587; Bettman 326, 328, 332, 393, 432, 465, 483, 590, 608, 611; Dean Conger 484; Corbis 561, 562, 565; Paul Hardy 569, 575; Liz Mangelsdorf/San Francisco Chronicle 355, 356; Robert Llewellyn 415; Francis G. Mayer 385, 387; Reuters 347, 435; Joseph Sohm/Visions of America 557, 567; Swim Ink 2, LLC 599; Ramin Talaie 350; Pennie Tweedie 529; **FONAR:** 410, 412, 413, 417; **Getty Images:** AFP/AFP 351, 381, 554; Agence France Presse/Hulton Archive 382; Robyn Beck/AFP 508; Blank Archives/Hulton Archive 329; Tim Boyle/Getty Images News 352, 556, 585; Matt Cardy/Getty Images News 489; Central Press/Hulton Archive 463; China Photos/Getty Images News 515; Rachel Cobb/AFP 482; Louis Daguerre/Time & Life Pictures 394; George Eastman House/Hulton Archive 459; Erris/Time & Life Pictures 581; Mehdi Fedouach/AFP 485; Andreas Feininger/Time & Life Pictures; Focus on Sport/Focus on Sport 631; Romeo Gacad/AFP 527, 530; Georges Gobet/AFP 383; Edward Gooch/Hulton Archive 388; David Greedy/Getty Images News 583; Henry Guttmann/Hulton Archive 600; Yvonne Hemsey/Getty Images News 477; Stan Honda/AFP 400; Sandy Huffaker/Getty Images News 371; Hulton Archive/Hulton Archive 369, 404, 454, 464, 466, 476, 538, 580, 589, 591, 593, 603, 609; Imagno/Hulton Archive 436; Yale Joel/Time & Life Pictures 624; Koichi Kamoshida/Getty Images News 519, 525; Keystone/Hulton Archive 472, 475, 478; 480, 551, 558; Toshifumi Kitamura/AFP 500; Jimin Lai/AFP 403, 405; Jerry Laizure/Getty Images News 419, 423; Michael Latz/AFP 407; Mauricio Lima/AFP 409, 414; Steve Liss/Time & Life Pictures 421; Doug McFadd/Getty Images News 367; David McNew/Getty Images News 431, 438, 486; MPI/Hulton Archive 471, 610, 616; John Phillips/Time & Life Pictures; Picture Post/Hulton Archive 539; Bill Pugliano/Getty Images News 602; Jaime Reina/AFP 517; Rischgitz/Hulton Archive 386, 545; Rosland Rahman/AFP 372; RDA/Hulton Archive 377, 378; Jean Baptiste Sabatier-Blot/Hulton Archive 390, 391; Eric Sander/3rd Party-Agents 373; Pascal Le Segretain/Getty Images Entertainment 401;Tim Sloan/AFP 447; Steven St. John/National Geographic 621; Stock Montage/Hulton Archive 572, 613; John Storey/Time & Life Pictures 493, 494; Justin Sullivan/Getty Images News 366; Mario Tama/Getty Images News 325, 615, 621; Three Lions/Hulton Archive 389, 406, 455; Time & Life Pictures/Time & Life Pictures 548, 578; Time Magazine/Time & Life Pictures 630; Topical Press Agency/Hulton Archive 597; USMC/Hulton Archive 330; Bruno Vincent/Getty Images News 449, 450, 451; Roger Viollet/Roger Viollet Collection 499; Werner Wolff/Time & Life Pictures 606; Yellow Dog Productions/The Image Bank 361, 363; **Quanah Gibson-Mount:** 528; **iStockphoto:** borsheim 577; eyecrave 428; Dave Gold 426; Grafissimo 547, 549; JenD 504; lisafx 399; lurii 617; photosbyjim 511; pichunter 623, 629; Rouzes 496; shooterguy 506; starfotograf 514; wachawachawant 425, 429; YinYang 570; **Jupiter Images:** 357; **Kodak:** 453, 460; **Library of Congress:** 354, 395, 420, 523, 566, 573, 604; **NASA:** 516; **Nobel Foundation:** 362, 365; **Palo Alto Research Center Inc.:** 375; **Red Cross:** 441, 442, 444; **Science Museum/Science and Society Picture Library:** 337, 342, 396, 457, 458, 461, 468, 470, 502, 503, 520, 521, 531, 532, 533, 537, 541, 542, 596; **United States Patent Office:** 359, 413, 422, 434, 457, 552, 628; **M. Weik:** 340; **Xerox:** 498